I0815390

THE COMPLETE GUIDE TO HOME PERMACULTURE

Quarto.com

First Published in 2026 by Cool Springs Press, an imprint of The Quarto Group,
100 Cummings Center, Suite 265-D, Beverly, MA 01915, USA.
T (978) 282-9590 F (978) 283-2742

EEA Representation, WTS Tax d.o.o.,
Žanova ulica 3, 4000 Kranj, Slovenia.
www.wts-tax.si

30 29 28 27 26 1 2 3 4 5

ISBN: 978-0-7603-9880-7

Digital edition published in 2026
eISBN: 978-0-7603-9881-4

Library of Congress Control Number: 2025944049

Design and page layout: Sarah Pyke
Cover Images: Erik Meadows
Illustration: Kate Freeman

Printed in Guangdong, China TT092025

THE COMPLETE GUIDE TO HOME PERMACULTURE

How to Transform Your Yard into a Thriving and Productive Ecosystem

BRANDY HALL of Shades of Green Permaculture

This outdoor shower reclaims water to passively irrigate planting areas.

CONTENTS

For Christopher

"*We are all just walking each other home.*"

—Ram Dass

And for my daughter, Zephyr Dove,

living proof that life is always longing for itself.

A gulf fritillary butterfly (*Agraulis vanillae*) drinks nectar from a blooming zinnia. This small moment shows how pollinators and plants support each other. In permaculture, we design gardens to create these kinds of helpful relationships right outside our door.

INTRODUCTION
THE THREE PILLARS OF A REGENERATIVE LANDSCAPE

As the founder and CEO of Shades of Green Permaculture, I've spent most of the last two decades studying and refining my understanding of how permaculture fits into modern urban life. Since I began the company in 2008, I have refined the ways we approach permaculture in an urban/suburban context, within a "City in the Forest," as Atlanta is affectionately known. Atlanta is a temperate climate city of over five million people, situated in the piedmont of the southern tip of the Appalachian Mountains, in the southeast United States.

At the time I'm writing this book, Shades of Green Permaculture has grown to a twenty-five-person staff, and we have worked with thousands of clients and students to design and implement regenerative landscapes at homes, schools, businesses, new developments, city parks, and more. Over the years, I've synthesized our work, including the trials, errors, case studies, and key learnings, into a simplified and easy-to-implement framework I call the Three Pillars of a Regenerative Landscape.

Permaculture can be a daunting and convoluted subject that people have a hard time defining, much less practicing. Despite many claims that permaculture is complicated or dogmatic, it's actually quite simple when we think of our land management strategies in these three distinct yet interconnected areas of influence: restoring the water cycle, building soil fertility, and cultivating diverse plant communities that protect biodiversity while growing food, medicine, and pollinator habitat.

I've heard from countless sub/urban residents who believe they can't practice permaculture unless they're growing most of their own food. I've also seen those beliefs change when people realize permaculture looks different depending on your site context and goals. City dwellers are contending with a loss of topsoil, extreme stormwater runoff from rampant development, and aggressive species crowding out biodiversity. Not to mention less-than-ideal growing conditions, homeowner's association (HOA) constraints, city ordinances, and exploding populations. In fact, an estimated 68 percent of the world's population are expected to live in urban centers by the year 2050.

In the city, our clients are drawn to the idea of growing food but feel constrained by the daily grind of forty-hour work weeks, hours-long commutes, and all of the other obligations that take a city dweller off-site. The goal for the majority of our clients is not to grow all of their own food, but rather to maintain a thread of connection to the

Opposite, top left My daughter, Zephyr, grazing kale from our garden. Witnessing her grow up in an environment that's healthy and organic, where she feels at home with the plants and wildlife, is one of my greatest joys and a big reason I do this work.

Opposite, top right Echinacea (*Echinacea purpurea*), muhly grass (*Muhlenbergia capillaris*) and Russian sage (*Salvia yangii*) thrive beneath a canopy of edible serviceberry (*Amelanchier canadensis*) in the unused strip between the road and sidewalk in a city. Every space has the potential to hum with life!

Opposite, bottom left Echinacea (*Echinacea purpurea*), muhly grass (*Muhlenbergia capillaris*), Russian sage (*Salvia yangii*) and serviceberry (*Amelanchier canadensis*) make a low-maintenance plant palette for any area with notoriously difficult growing conditions. With minimal establishment irrigation, this plant community provides year-round seasonal interest and much-needed pollinator habitat.

Opposite, bottom right My daughter excitedly harvesting serviceberry (*Amelanchier canadensis*) in late spring. Edible fruits planted in rights-of-way make walks more fun.

natural world and make sustainable land management decisions that align with their ethics—be it mowing with electric equipment, forgoing herbicides and pesticides, or reducing their lawns in favor of more native and useful plants.

I strongly believe that we don't need to live off-grid, growing all of our own food, being "self-sustaining" (what organism on Earth is "self-sustaining" anyway?), quietly going about our business on a large private acreage of land. I believe it is possible to have a life-giving, reciprocal relationship with land that supports us, even in the urban/suburban context. By creating vibrant landscapes that build health rather than undermine it, we have the tools to address the climate crisis in our yards because small solutions equal big impact, and it adds up quickly.

Below Even tiny yards in urban areas can vibrate with life. Native plants, fertile soil, and pollinators abound in this pocket-sized herb garden.

My Big Why

I came to this work in a circuitous way, heavily informed by my experiences as a child growing up between my mom and stepdad's conventional plant nursery in South Florida and my dad's home in the pristine western North Carolina mountains. When I

Above, left Muscadine (*Vitis rotundifolia*) growing on a trellis in an inner-city backyard. Perennial edible cutleaf coneflower (*Rudbeckia laciniata*) naturalizes in the foreground. A tunnel trellis covers a small in-ground bed that grows summer squash. Small spaces can grow an abundance of food by layering vertically.

Above, middle My daughter playing on mountains of organic compost instead of chemical fertilizers, like I did as a child.

Above, right Old photos of the nursery in South Florida, circa 1981. Drainage canals bisect the farms; pesticide and herbicide runoff make the water toxic.

was young, my mom and stepdad, who ran a plant nursery in South Florida in the 1980s, were poisoned by the herbicides and pesticides they used on the farm, which led to escalating allergic reactions and long-term impacts on their health.

As a child, I spent my days climbing piles of soil and fertilizer, imagining them as mountains in my own little world. I ran barefoot through the greenhouse drainage ditches, stopping to watch pincher bugs battle it out in the muddy water. What I didn't realize then—and what my parents wouldn't fully understand until years later—was that the pesticides saturating our surroundings were seeping into the groundwater, soaking into the soil, and accumulating in their bodies. In Palm Beach County, where we were surrounded by tomato fields, nurseries, and the constant hum of crop dusters overhead, pesticide exposure wasn't just likely—it was unavoidable.

At the time, my parents were regularly spraying diazinon, a pesticide that was once common in households but later became a Restricted Use Pesticide under regulation of the Environmental Protection Agency (EPA). They followed the required safety protocols, wearing the protective suits designed to keep them safe. But in the sweltering South Florida summers, those suits quickly became drenched in sweat, turning them into a trap rather than a shield. Instead of keeping the chemicals out, the moisture inside held them against their skin, allowing them to seep into their bodies. My mom often described a strange metallic taste in her mouth every time she sprayed. Over time, both she and my stepdad developed worsening reactions to pesticide exposure—effects that didn't fade even after they stopped using chemicals in their own nursery.

Then, in August 1996, my stepdad's health took a sharp turn for the worse. He developed a mysterious illness that doctors struggled to diagnose. Some thought it was lupus, others suspected gout. Some diagnosed rheumatoid arthritis, while others feared cancer. All they knew for sure was that he couldn't walk—his ankles, knees, and hips were so swollen that he needed them drained multiple times a week. Uncontrollable nosebleeds would last for up to an hour, and for nearly a year, he was confined to a wheelchair. Doctors doubted he would ever walk again. The cause, however, was clear—earlier that day, he had delivered seeds to a nursery that had just been sprayed with pesticides. Hours later, his joints were swelling, and his body was shutting down.

My mom's reactions grew worse, too. At first, her tongue would swell slightly after exposure, and she could manage it with Benadryl. But over time, she no longer had to handle the chemicals directly in order to have a reaction—just driving past a freshly sprayed field was enough to send her into anaphylactic shock. The fly-over mosquito spraying, common in Palm Beach County, or even a neighbor's termite fumigation could trigger a reaction. Our family was placed on the county's "chemical sensitivity" list, which meant we'd get notified before any aerial spraying. At least twice a month, we'd pack our bags, flee to another county, and stay in a hotel until it was safe to return home.

This routine continued for years while my parents remained in the nursery business. My mom took over operations while my stepdad was still in a wheelchair. But one day, a simple delivery nearly ended in tragedy. As she pulled up to a farm, she caught the unmistakable scent of recently sprayed pesticides. Acting quickly, she rolled up the windows and sped away. Within an hour—despite three doses of Benadryl—her tongue had swollen so much that she was choking on it. Her body convulsed violently, shaking the car door panels loose as we raced to the hospital. By the time we arrived, her blood pressure was so high that doctors said she was on the verge of a stroke.

That moment was the breaking point. A week later, my parents packed up our entire life and moved to the woods of Columbia County, Florida, where they would spend the next four years detoxing.

Below, left A honeybee sips the nectar of lavender (*Lavandula angustifolia*). Planting pollinator habitat is one of the many ways we can restore connection and bring fertility back to the land.

Below, right Historical texts from the late 1800s and early 1900s describe the relationship between passenger pigeons (*Ectopistes migratorius*) and the American chestnut (*Castanea dentata*) that once dominated the deciduous hardwood forests of the eastern United States.

Later, a hair analysis confirmed what they had suspected—their bodies were saturated with aluminum and mercury. Even years after exposure, simple things like the scent of nail polish could set off an allergic reaction. They completely overhauled their lifestyle—switching to organic food, eliminating toxic household products, and banning conventional beauty products in our home. Over time, their bodies slowly healed. Eventually, they regained enough health to rejoin society and live with fewer restrictions—but, as you might imagine, as a young person this made a huge impact on my view of the world and left me searching for a better way.

The other thread of my life takes place in western North Carolina, the place of my birth, the home of my father and my ancestors dating back to the mid-1600s. As with two very different sides of a coin, my childhood in southern Appalachia was punctuated with pristine lakes, seemingly endless temperate rainforest, wagon trains, fireflies, and other idyllic scenes. I spent half of my childhood swimming in clear mountain lakes, riding horseback through tall meadows, drinking spring water straight from the mountain, and eating fresh from the garden. These experiences led to my work—as builder, creator, and educator—and a lifelong desire to create harmony between the human-built environment and the natural world.

This juxtaposition laid the foundation for a lifelong inquiry: Are our actions as humans innately destructive and extractive, or can we coexist with our environments in a way that our actions are actually *restorative*? Rather than always taking, can we become so integrated with our landscapes, despite modernity, that our outputs feed a system larger than ourselves? Can we feed in order to be fed, rather than the other way around?

Let me illustrate these questions with an analogy.

In the late 1800s, the American South was full of the now-extinct passenger pigeon, *Ectopistes migratorius*. They were so prevalent, in fact, that in his account from his 1835 book, *Ornithological Biography or An Account of the Habits of the Birds of the United States of America*, John James Audubon describes the sky darkening with clouds of pigeons, lasting three days:

> "The air was literally filled with Pigeons; the light of noon-day was obscured as by an eclipse; the dung fell in spots, not unlike melting flakes of snow, and the continued buzz of wings had a tendency to lull my senses to repose ... I cannot describe to you the extreme beauty of their aerial evolutions, when a hawk chanced to press upon the rear of the flock. At once, like a torrent, and with a noise like thunder, they rushed into a compact mass, pressing upon each other towards the center. In these almost solid masses, they darted forward in undulating and angular lines, descended and swept close over the earth with inconceivable velocity, mounted perpendicularly so as to resemble a vast column, and, when high, were seen wheeling and twisting within their continued lines, which then resembled the coils of a gigantic serpent ... Before sunset I reached Louisville, distant from Hardensburgh fifty-five miles (89 km). The Pigeons were still passing in undiminished numbers and continued to do so for three days in succession."

These birds were more than just an awe-inspiring spectacle; they were an essential part of a finely tuned ecological cycle. The passenger pigeons migrated from their wintering grounds in the phosphorus-rich estuaries of the Gulf of Mexico back to the deciduous hardwood forests of the southern Appalachian Mountains, where they feasted on the abundant nuts of the American chestnut (*Castanea dentata*). These towering trees, which once made up nearly 25 percent of the eastern forests in the United States, produced enormous quantities of food, supporting not just pigeons but bears, deer, and countless other species. Appalachian lore tells of chestnut trees so full of pigeons feasting that there were mounds of guano as tall as a grown man beneath each tree.

The pigeons, in return, enriched the soil with phosphorus from their droppings—nutrients transported from coastal estuaries back to the mountains, completing a cycle that had been refined over millions of years. This exchange sustained the health of the chestnut forests, which, in turn, provided habitat and food for an entire web of life. The chestnuts fed the pigeons, the pigeons fertilized the soil, and the forest thrived. It was a perfect example of ecological intelligence: a system in which each participant contributed to the health of the whole. The chestnut trees, in their prolific fruiting, attracted the necessary nutrients to their community so that the entire ecosystem could thrive. They were feeding, in order to be fed.

But then, we broke the cycle.

In a matter of decades, human interference wiped out both species. The passenger pigeons were hunted to extinction by the early 1900s, victims of relentless commercial hunting and habitat destruction. Without their migrations, the natural phosphorus cycle that had sustained Appalachian forests for thousands of years collapsed. Then, in the early twentieth century, the chestnut blight—introduced by imported trees carrying the invasive *Cryphonectria parasitica* fungus—swept through the eastern United States, decimating nearly every mature American chestnut in its path. And the ones that might have survived were cut down because of short-sighted thinking and a desire to get what was left, before it dies. Unfortunately, logging the remaining trees removed almost any remaining resistance against the chestnut blight from the gene pool. In less than a century, both the passenger pigeon and the American chestnut were gone, their deeply intertwined ecological roles erased.

And this wasn't an isolated incident. This is what happens when we ignore the intelligence of nature—when we extract, disrupt, and take without giving back. It's what's happening now, in a million different ways. When we pave over wetlands, we disrupt natural flood control. When we poison pollinators with pesticides, we weaken the very systems that allow food to grow. When we remove native plant communities for monoculture lawns, we sever the intricate relationships between soil, water, and wildlife.

But here's the hopeful part: We don't have to keep repeating these mistakes.

By adopting the practices in this book, you are helping to restore the natural cycles that human interference has broken. When you plant native trees and diversify your landscape, you reestablish lost connections. When you build soil instead of depleting it, you bring fertility back to the land. When you harvest rainwater to keep it on-site and restore habitat, you become an active participant in regeneration, rather than destruction.

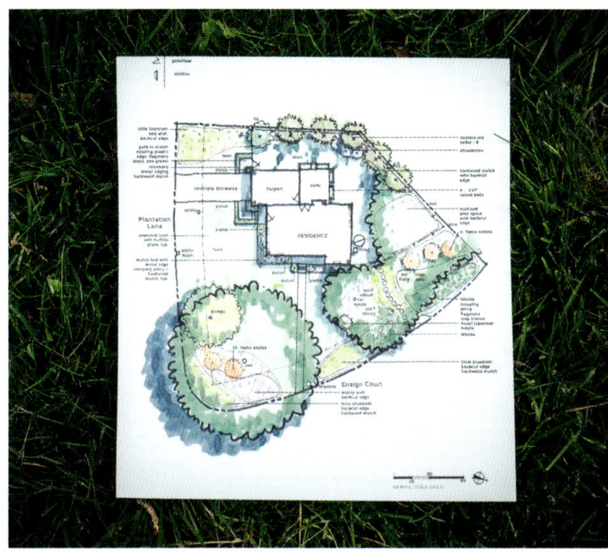

Above, left My daughter harvesting wineberry (*Rubus phoenicolasius*) from our patch. This is an aggressive species in some climates, but not in Atlanta because it gets too hot. For more discussion about choosing appropriate plants, see chapter 8.

Above, middle Native wild hydrangea (*Hydrangea arborescens*) thriving in a garden without irrigation, despite an extended drought and unseasonably hot summers. Planting native and appropriate species on the edges of their ranges and with dense coverage of the soil helps mitigate the effects of drought.

Above, right Permaculture landscaping integrates food production, rainwater harvesting, and native plantings, balancing human use and ecological health.

The chestnut and the passenger pigeon are gone, but their lesson remains: We can either take from the land until there is nothing left, or we can work with nature, understanding its intelligence and helping it to heal, and perhaps healing ourselves in the process. The choice is ours, and it starts in our own backyards. Our landscapes, and our actions as land stewards, hold this same potential.

What Is Permaculture?

The word *permaculture* is a portmanteau of the words "permanent" and "agriculture" that points to creating lasting and productive systems that yield food, fiber, and fuel year over year. Permaculture is fundamentally an approach to design that prioritizes ethical principles and ecological harmony. It focuses on blending human activity with natural systems to establish resilient, self-sustaining environments. While permaculture principles can be applied to a wide range of systems—both land-based and beyond—the specific methods used will vary depending on the unique conditions and needs of each setting.

For the purposes of this book, I will teach you a simple and effective approach for applying permaculture to the residential landscape in sub/urban settings. In a city, the way we apply permaculture may look different from how it is applied on a rural farm. For that reason, it is important to first understand the patterns that form the foundation of regenerative human development, and from there we can make incremental changes in our own landscapes that mimic natural systems and help forward the work of Earth repair.

You'll find me using *regenerative* and *permaculture* interchangeably throughout this book. I do this intentionally, as my understanding of permaculture has expanded beyond the food-production-centric thinking of previous decades. At this point, I consider any

Top left A typical suburban client's unused lawn space before planting.

Left Immediately after pulling up the lawn, installing rain gardens, and planting perennials. The gardens take time to establish, but planting smaller plants and allowing for gradual establishment yields a better product in the end.

Top right The garden beds didn't look like much at first, but after one growing season, the meadow species are flourishing with minimal maintenance and no irrigation except what passively feeds the rain gardens below.

Right A garden bed full of herbs and cut flowers. Creating beautiful and productive spaces in our yards encourages us to become both benefactors and beneficiaries of the Earth's abundance.

regenerative activity we do as land stewards to serve the same purpose: reconnect us as collaborators with the landscapes we dwell within. Whether we're growing food for ourselves, growing our landscape's capacity to retain water, growing habitat for wildlife, growing soil microbiota for healthy plants—or ideally some combination of all four of these—we are both benefactors and beneficiaries of the Earth's abundance and constant movement toward balance and health.

As a permaculturist working in a largely human-centric, developed city, I see my work as a dance between the built and the wild, weaving them into a conversation where each enriches the other, bringing life to the spaces where they meet. So much of urban development is about carving spaces, manipulating topographies, defining borders, and taming the wild. To add another layer of complexity to the issue of land management, we live in a time when climate extremes are becoming normal. "Record-breaking" and "unprecedented" seem to precede every weather-related headline: temperatures, snowfall, wildfires, rain events, hurricane strengths. This year alone, I can recall several "record-breaking" events: Hurricane Helene traveling nearly six hundred miles (966 km) inland and delivering a thirty-thousand year flood event to Asheville, North Carolina; eleven inches (28 cm) of snowfall in Louisiana; multiple catastrophic wildfires raging through Los Angeles, California, simultaneously; "unprecedented" low water levels in Lake Mead and Lake Powell in the Southwest United States.

I don't say all of this to incite fear or hopelessness. Quite the contrary. We have the tools to make small changes in the landscapes we steward that amount to huge impacts on the broader landscapes we depend on. I've seen it in my own work in Atlanta, where one resident builds a beautiful and resilient ecological landscape, and others, in their inspiration, then follow suit. Soon, an HOA is rewriting their bylaws and new developments are embracing the principles of ecological design at the outset.

Regenerative landscaping is the antithesis of instant gratification. Instead of rushing to create a "finished" garden in a weekend or season, you're invited to engage with your land over time. This approach values long-term sustainability, natural processes, and thoughtful design choices that reflect the unique qualities of our environments. You're encouraged to move away from the mentality of instant green lawns or quickly assembled ornamental flower beds and instead focus on nurturing regenerative and resilient ecosystems. In a world where our relationship to the land is increasingly important, transforming a conventional landscape into a beneficial ecosystem is a beautiful endeavor. Whether you have a sprawling backyard or a smaller space, there are ways to create a productive and eco-friendly permaculture garden that provides food, habitat for wildlife, and a sense of connection to the natural world.

In *The Complete Guide to Home Permaculture*, you will understand permaculture in simple, actionable concepts through case studies and techniques adapted to the sub/urban landscape and modern lifestyle. I'll walk you through the uncomplicated, tried-and-tested process I've used with clients over two decades, showing you how to transition conventional landscapes into beneficial and interconnected ecosystems. These landscapes help restore the water cycle, capture carbon, protect biodiversity, and grow food, medicine, and habitat. *The Complete Guide to Home Permaculture* offers solutions to the global climate crisis we face, where every day, urban dwellers can participate in the Earth's return to health in their own backyards.

I offer this book to you as a guide into regenerative land stewardship, so you can craft a beneficial ecosystem in your yard, that it may feed the more-than-human world and a time beyond our own.

This book is organized into four sections, each designed to guide you through the essential principles and practical applications of home permaculture. The first three sections align with the Three Pillars of a Regenerative Landscape, which serve as the foundation for creating a thriving, self-sustaining yard. Section 1 focuses on Water, exploring how to restore the natural water cycle, reduce irrigation needs, and implement techniques like swales and rain gardens to retain moisture in the soil. Section 2 delves into Soil, highlighting the importance of building soil fertility through composting, mulching, and regenerative gardening techniques that nourish the earth instead of depleting it. Section 3 turns to Plants, showing how to design diverse plant communities that protect biodiversity, feed pollinators, and grow food while maintaining ecological balance.

But theory alone isn't enough. That's why section 4 brings these principles to life through real-world case studies. This section showcases actual properties—just like yours—where these strategies have been put into practice. You'll see how a small urban yard was transformed into a resilient oasis, how an herbalist's home apothecary was designed for both beauty and function, and how even a shady backyard can become a productive food garden. These stories prove that permaculture is not just an abstract concept—it's something you can implement in your own space, regardless of its size or challenges. As you move through this book, let these examples inspire you to take action, knowing that every step you take toward regenerating your landscape contributes to a healthier planet.

Now that we've laid the foundation for what permaculture is and how it can be applied in urban and suburban settings, it's time to dive into the Three Pillars of a Regenerative Landscape, and it all begins with water. Water is the most essential element of any ecosystem, yet conventional stormwater management treats it as a nuisance to be drained away as quickly as possible. In section 1, we'll explore the problems with this approach and discuss strategies to restore the natural water cycle—slowing, sinking, and spreading water where it's needed most.

Gulf fritillaries flock to nectar-rich zinnias, bringing beauty and vital pollination to your veggie garden. Their presence boosts crop yields and supports a balanced, biodiverse ecosystem.

A 300-gallon (1,136 L) rainwater collection cistern is one technique to manage water as a resource in your garden. The stored water can be used for many purposes, and the overflow can be directed strategically through the yard to passively irrigate planting areas.

Section 1

Water

Pillar #1: Manage Water as a Resource to Rehydrate Soils, Reduce Irrigation, and Restore the Water Cycle

Water is the foundation of life. It shapes landscapes, nourishes ecosystems, and influences weather patterns. Yet, in modern landscapes, we often treat water as an afterthought—something to be drained away as quickly as possible rather than as a resource to be carefully stewarded. Our conventional approach to water management prioritizes convenience over resilience, leading to flash floods, droughts, and depleted aquifers. But what if we could work with water instead of fighting against it?

In this section, we'll explore how to restore the natural water cycle in our own yards—rethinking stormwater management, designing landscapes that retain moisture, and using water wisely to create thriving ecosystems. You'll learn why conventional water practices fail and how simple techniques like rain gardens, swales, and rainwater harvesting can turn excess runoff into an asset, helping to rehydrate the soil, reduce the need for irrigation, and prevent erosion.

By the end of this section, you'll see water differently—not as a problem to be managed, but as a precious, abundant force that brings fertility and resilience to your landscape. Whether you live in a rainy climate or an arid region, you'll discover practical, proven ways to harness water's natural rhythms, ensuring that every drop is put to work in service of a healthier, more self-sustaining garden.

"Where water runs, make it walk."

MASANOBU FUKUOKA

When we don't manage water as a resource, it can be a destructive force. Runoff from storms that cannot sink into the soil means that flash flooding is more likely. The city of São Leopoldo, Brazil, was devastated by flooding in the spring of 2024.

CHAPTER 1

THE PROBLEM WITH CONVENTIONAL WATER MANAGEMENT

How is water usually managed? Water management, as conventionally practiced, is barely managed at all. When it rains, the prevailing approach treats water as a nuisance to be removed as quickly as possible. Instead of seeing rain as a resource, conventional systems funnel it into stormwater drains, carrying runoff from impervious surfaces like roads, rooftops, and parking lots into our rivers, streams, and waterways.

Removing trees—a common practice in urban and suburban development—further exacerbates the problem. Without trees to intercept and absorb rainfall, water rushes into streams rather than soaking into the soil, intensifying issues like erosion and flooding. Over time, this contributes to the urban heat island effect, where concrete and asphalt amplify temperatures. The increased heat further disrupts local rainfall patterns, leading to long-term desertification.

Simply put, more heat means less rain, and less rain leads to degraded ecosystems.

Before we seek solutions, it is imperative to identify the problem correctly. In this chapter, we'll investigate the status quo with conventional urban stormwater management, how that contributes to a broken water cycle, and how that, in turn, affects climate change. There are many seemingly innocuous water practices in home landscapes that actually compound bigger issues like drought, depleting aquifers, and waterway sedimentation regionally, nationally, and globally. Understanding these detrimental practices provides an opportunity to restore the water cycle, beginning with our own yards.

Understanding the Water Cycle

The global water cycle is akin to the human body's circulatory system, and understanding the movement of water on the planet gives insight into the incredible importance water plays in global heat dynamics. It is estimated that between 75 percent and 95 percent of global heat dynamics are actually regulated by water, whereas between 4 percent and 11 percent are regulated by carbon. With the focus zeroed in on carbon emissions in the fight against global warming, we may be missing the forest for the trees.

Water circulates throughout the Earth in a process called the hydrologic cycle. In an undisturbed scenario,

their site work costs. A humus-rich, spongy topsoil layer helps the ground hold water, and the porous quality of topsoil allows more water to soak in, rather than run off.

COMPACTING THE GROUND

Scraping the topsoil exposes the subsoil. During construction, in addition to staging materials and machines, site grading compacts subsoil even further in order to finesse changing topographies. When compacting the ground for stabilization, especially when combined with other practices like removing trees and scraping topsoil, porous soil diminishes, and site conditions prohibit water from sinking into the soil. Rather than being absorbed into the ground, water from rooftops and surface runoff is channeled away from the site, flowing into streams and waterways while dragging sediment and soil along with it. Have you ever noticed the murky color of rivers after a heavy rain? This is generally caused by destabilized and exposed soil. This sediment-laden runoff contributes to water pollution and less sequestration of carbon dioxide (CO_2) in the soil.

IRRIGATING FROM DEPLETED SOURCES

When irrigation pulls from city water, wells, rivers, or other sources, while simultaneously sending rain "away" via stormwater infrastructure, it creates linear movement where the water is not cycled back into the system, so it doesn't replenish itself. In this case, groundwater and aquifer recharge diminishes, lake levels drop to historic lows, and river basins dry up. According to the United States Environmental Protection Agency (EPA), 30 and 60 percent of residential water consumption in the United States is for watering lawns, and to exacerbate the issue even further, "some experts estimate that as much as 50 percent of water used for irrigation is wasted due to evaporation, wind, or runoff caused by inefficient irrigation methods and systems."

The impacts of evaporation—accelerated by deforestation or water runoff rather than absorption—ultimately affect other parts of the Earth, disturbing the hydrologic cycle and creating atypical weather events. Furthermore, less carbon in the soil combined with fewer trees, means less water is stored in the ground, which creates a "heat island effect," leading to desertification over time. More heat means more evaporation, which means less water vapor emitted over time, which adds up to less rain.

Peter Eagleson, in his article "The Role of Water in Climate," explains the dynamic nature of global water circulation, stating that "water is not just a passenger on the passing breeze, but to a large extent creates the breeze that transports it." He further emphasizes the interconnectedness of this system, noting that changes in evaporation in one region can influence precipitation patterns elsewhere. As an example,

he describes how moisture evaporated from the forests of Southeast Asia eventually falls as rain over eastern China and much of the United States.

In the next chapter, I will discuss the first pillar of a regenerative landscape, and the different strategies you can employ to restore the water cycle in your own yard.

Conventional water management systems prioritize removal over retention, leading to a host of ecological problems, from flash flooding to drought. In chapter 2, we'll shift our focus from identifying these problems to implementing solutions. By recognizing the signs of a broken water cycle in your own yard, you can begin making changes that capture and use water wisely—restoring natural hydrological balance and creating a more resilient landscape.

Below Our reservoirs are increasingly drying up. Using city water to irrigate exacerbates this problem.

Create opportunities for water to sink into the soil. Using permeable surfaces like this flagstone path with aggregate infilling the joints lets the water infiltrate rather than run off, the way it would on concrete.

CHAPTER 2

HOW TO RESTORE A BROKEN WATER CYCLE

Now that you understand the causes and implications of a broken water cycle, I want to discuss the first pillar of a regenerative landscape: Manage water as a resource to rehydrate soils, reduce irrigation, and restore the water cycle. In this chapter, you will identify common site conditions that indicate an opportunity to better manage water. By reading existing site conditions, you will be able to correctly diagnose the water issues in your yard and understand which water management techniques will be most applicable to your site moving forward.

When treated as such, water is precious, and water is abundant.

Healthy Water Cycles, Healthy Gardens

In a forest, rain forms a mist as it's broken up by plant leaves while falling from the sky. The mist falls slowly and 80 percent of the water sinks into the soft, leaf litter-covered ground. Forest soils store massive amounts of water, making it available to trees throughout the year. A healthy water cycle is one in which water moves through the environment in a balanced, self-sustaining way, supporting ecosystems, recharging groundwater, and maintaining climate stability. In natural systems, rain is absorbed by healthy soil, percolates into aquifers, flows gradually into streams and rivers, and eventually returns to the atmosphere through evaporation and transpiration. This process keeps landscapes hydrated, prevents flooding, and ensures water availability even during dry periods.

In contrast, when the water cycle is disrupted—often due to deforestation, urbanization, and poor land management—rainwater runs off too quickly, leading to erosion, drought, and declining water tables. Healthy ecosystems, especially forests, wetlands, and grasslands, act as sponges that retain moisture, slow water movement, and replenish underground reservoirs.

For permaculturists, restoring a healthy water cycle means designing landscapes that slow, spread, and sink water instead of allowing it to run off. This involves using techniques like rainwater harvesting, swales, and rain gardens. When the water cycle functions properly, it supports abundant plant and animal life, stabilizes local climates, and keeps ecosystems thriving.

It All Starts with Water

The foundation of any landscape is water. Where there is water, life quickly follows.

When you're starting to build your permaculture landscape, you will begin with water. We can look to nature's quintessential permaculturist, the beaver, to understand the impact of managing water as a resource.

Beavers are nature's master water engineers, transforming dry, degraded landscapes into lush, thriving ecosystems. Their dams slow down the flow of water, allowing it to spread across the land instead of rushing away. This process recharges groundwater, keeps streams flowing during droughts, and creates wetlands that support an incredible diversity of life. By holding water in the landscape, beavers effectively prevent erosion, store moisture in the soil, and stabilize entire ecosystems.

As permaculturists, we can take inspiration from their approach to water management. Instead of letting rainwater drain away, we can design landscapes that

Below An aerial view of a client's shade garden. All paths are gravel, which allows the water to sink in. Any area that isn't a permeable path in this yard is instead growing a dense ground cover layer, which helps keep the soil moist and protects tender plant roots from drought.

Right Beavers are nature's master water engineers, transforming dry, degraded landscapes into lush, thriving ecosystems. Their dams slow down the flow of water, allowing it to spread across the land instead of rushing away.

capture, store, and redistribute it, much like beaver ponds do. Techniques such as swales, rain gardens, and natural wetland restoration mimic their impact, creating resilient systems that remain hydrated even in dry seasons. Beavers instinctively understand that water is life, and by thinking like them, we can restore landscapes, enhance biodiversity, and build climate resilience. They don't just survive in their environment—they regenerate it, and that's a model worth following.

Managing water is the foundation of an ecology where nature thrives. In a regenerative landscape, rain gardens concentrate water from downspouts into basins in the yard, which allows infiltration into the ground, and reduces runoff, flash flooding, drought, and city water costs. A hydrated soil is a healthy soil. Healthy soil grows healthy plants. Healthy plants create healthy yards and gardens, which help create healthy ecosystems full of butterflies, beetles, songbirds, bees, fruit trees, nut trees, berry bushes, cut flowers, perennial vegetables, and culinary and medicinal herbs.

The Water in Your Landscape

Identifying the various sources of water on your site is essential when creating a permaculture design because water is the foundation of all life and productivity in a landscape. Understanding where water comes from, how it moves, and how it interacts with the land allows for strategic planning to maximize its benefits while minimizing waste and erosion.

WHAT ARE THE SOURCES OF WATER ON YOUR SITE?

Each source—rainfall, groundwater, surface water (rivers, ponds, springs), and even human-made sources like greywater or runoff from roofs—has characteristics that influence how it should be captured, stored, and used. By mapping these sources, you can design systems that slow, spread, and sink water into the land, ensuring that it stays available for plants, animals, and soil life even during dry periods. I like to think of this as "dancing the water through the landscape," employing different techniques that respond to your unique site conditions, so as to move the water elegantly throughout the site.

For example, if your site has a seasonal seepage, you might integrate small check dams or retention ponds to slow the flow and recharge groundwater. If rainfall is your primary source, you would focus on capturing it through swales, rain gardens, and tanks. Identifying high and low points in the landscape helps determine natural drainage patterns, allowing you to position plantings, water storage, and infrastructure accordingly.

EXISTING WATER CONDITIONS

Observing water movement on a site is crucial for effective permaculture design. Different water issues—such as wet or dry areas, erosion, and stormwater flow—indicate how water circulates and interacts with the landscape. Identifying conditions like pooling, runoff patterns, seepages, and existing drainage infrastructure helps determine where water can be captured, stored, or redirected. By mapping these elements, you can create strategies that enhance water retention, prevent erosion, and solve drainage issues.

Observe how each of these sources circulate in and on your site. Refer to the table below.

SOURCES OF ON-SITE WATER	
Rain	Includes what falls on a site and what runs onto a site from higher on the watershed as stormwater
City water	Water that comes from taps and spigots conveyed from municipal water supply
Wells	Water that comes from taps and spigots conveyed from dug or bored holes on a site
Springs	Tapped and untapped sources of groundwater that surface from below ground
Ponds and lakes	A natural or constructed body of still water
Creeks, streams, and rivers	Moving streams of water of varying sizes that convey a combination of groundwater from springs and rainfall

ASKING THE RIGHT QUESTIONS

Finding the right solutions begins with asking the right questions. When you're reading your site to determine which water management strategy you need, your observations will lead to the right questions.

Is Too Much Water the Issue?

If so, this would be indicated by conditions such as soggy ground, pooling/standing water, or flooding basements and crawl spaces. If you determine too much water is an issue, the next question to ask is "Where is the water coming from?" Go higher on the watershed, meaning uphill, to investigate. The likely culprits are your downspouts, your neighbor's downspouts, ineffective grading, and/or hardscaped surfaces.

Is Too Little Water the Issue?

If so, this would be indicated by conditions such as erosion, exposed subsoil, lots of impervious surfaces, or soil compaction. If you determine too little water is an issue, the next question to ask is "Where is the water going?" The likely culprits are fast-moving water actively leaving your site, rather than sinking into the soil; impervious surfaces that drain into stormwater infrastructure; or compacted soil and turf not allowing for infiltration. Did you know that as much as 85 percent of water that hits turfgrass can run off as stormwater?

You will likely have a combination of wet and dry issues, and during your observations, you will begin to see how water behaves on your site. Once you understand how water is moving through your site, you can next figure out which strategy, or combination of strategies, will be the most helpful to direct the water where it's useful. With these strategies, you'll be able to absorb water where it's needed, and remove it from where it's problematic.

MAPPING THE WATER IN YOUR LANDSCAPE	
Wet areas	Includes low spots and areas of pooling during heavy rains. Identify the nature of the water: Is it constant? Is it during a particular time of year or season? Can you tell where it comes from?
Dry areas	Includes beneath dense tree canopies, along slopes, near the thermal masses of concrete, walls, driveways, and roads
Visible erosion	Marked by exposed subsoil, visible tree roots, or gullies and furrows
Bodies of water	Includes floodplain or floodway limits and stream setbacks as determined by your county or municipality
Springs and seepages	Visible heads of waterways or consistently soggy ground marked by water-loving vegetation such as sedges or rushes
Stormwater infrastructure	Includes concrete channels, storm drains, headwalls, and drainage easements
Downspouts	Includes the approximate area of the roof they capture and where they are directed
Spigots and yard hydrants	Includes both potable and non-potable water sources
Drainage infrastructure	Locations and sizes of drain grates and pipes (if known) and amount of watershed captured

UNDERSTANDING WATER'S BEHAVIOR: THE KEY TO EFFECTIVE DESIGN

Water is one of the most powerful forces shaping our landscapes. It carves canyons, nourishes ecosystems, and when left unmanaged, can also erode soil, cause flooding, and deplete resources. But despite its immense influence, water follows six simple behaviors—patterns that govern how it moves, interacts with the land, and either regenerates or depletes an environment. When we understand these behaviors, we can design systems that work with water rather than against it.

Many land management challenges—whether it's erosion, drought, poor soil fertility, or flooding—stem from a misunderstanding of how water naturally behaves. Rather than forcing water to follow unnatural paths, successful design harnesses its tendencies, slowing it down when needed, redirecting it efficiently, and ensuring that it nourishes the land instead of stripping it away.

Whether you're designing earthworks, rainwater harvesting systems, or irrigation systems, the ability to "read" water's movement is the foundation of effective, long-term water management.

By understanding the six fundamental behaviors of water, you'll be equipped to implement techniques that capture, store, and utilize water wisely, creating landscapes that thrive in both wet and dry conditions.

Far left Water seeks its own level, as demonstrated by this rudimentary water-leveling experiment.

Middle Water moves swiftly in the narrow passages of mountain rivers, running down steep slopes through canyons.

Left Water moves much slower when it has space to expand.

1 Water is lazy

It flows downhill and always seeks the path of least resistance.

2 Water finds its own level

Let's consider water's leveling behavior with a simple activity. If you fill a clear tube with water and hold both ends evenly, the water is level with itself. If you lower one end, water seeks its level and flows out of the other end.

3 Water conveys nutrients

Water carries nutrients with it as it flows over and through soil.

4 Water moves faster when it constricts

Imagine the high-walled canyons of the glacial rivers of the Pacific Northwest, or the steep slopes of a rushing mountain river. Water moves faster through more narrow passageways.

5 Water moves slower as it expands

Picture the low-lying coastal rivers of the Southeast United States that slowly snake and wind their way through the coastal plains and seas of reeds. Water moves slower through wider passageways.

6 Water doesn't easily make 90-degree turns

Much like a merging acceleration lane on a freeway, quick-moving flows of water are more easily merged than when forced to make hard turns. Friction slows water down.

WATER IS ABUNDANT (EVEN IN DRY CLIMATES)

Even in the driest climates, water is far more abundant than we often realize. As previously mentioned, a single inch of rain on a 1,000-square-foot (93 sq m) surface can generate approximately 600 gallons (2,271 L) of water, and when we scale that up, the numbers become astonishing.

Consider an average-sized home in the United States, around 2,480 square feet (230 sq m). In a place like Atlanta, which receives an average of 53 inches (135 cm) of rain per year, that one roof alone could collect nearly 82,000 gallons (310,400 L) of water annually—more than enough to supply household and landscape needs with proper storage and management. Even in a much drier region that gets only 15 inches (38 cm) of rain per year, that same roof would still generate over 23,000 gallons (87,000 L)—a substantial amount when captured and used wisely.

This abundance of water is often overlooked because it quickly runs off, soaking into the ground or flowing into storm drains. But by understanding how much rainwater is actually available, we can shift from a mindset of scarcity to one of regenerative abundance, designing systems that capture and store this water where it's needed most. Whether in a wet or arid climate, recognizing rain as a resource—rather than something to be drained away—is the first step in creating landscapes that retain moisture, build resilience, and thrive in any conditions.

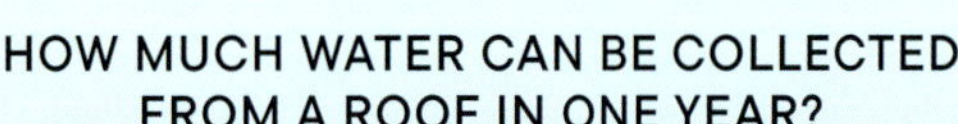

HOW MUCH WATER CAN BE COLLECTED FROM A ROOF IN ONE YEAR?

Use this equation to calculate the amount of rainwater that can be collected off the roof surface annually:

Area of a flat, impervious roof (sq ft) x annual rainfall (inches) x 0.622 = gallons of water generated annually

(Area of a flat, impervious roof [sq m] x annual rainfall [cm] x 10 = L of water generated annually)

Top Sunflowers are drought-tolerant powerhouses. As climate change brings hotter, drier conditions, plants like sunflowers play a vital role in creating resilient, low-water landscapes that still support pollinators and wildlife.

Bottom Russian sage (*Salvia yangii*), echinacea (*Echinacea purpurea*), and muhly grass (*Muhlenbergia capillaris*) are ideal meadow species for tough spots like the "strip of death" along roadsides. Drought-tolerant and low-maintenance, they thrive in harsh conditions while supporting pollinators and adding year-round beauty.

Opposite, top Even the smallest roofs yield high volumes of water over time. Capturing the water is a great way to have access to rainwater during dryer months.

Opposite, bottom Rain gardens help with drought by capturing and storing rainwater in the soil, keeping it available for plants during dry spells. They turn runoff into resilience—naturally.

RAINSAVER

Regenerative Water Strategies

The biggest impact in mitigating stormwater issues using the least amount of intervention happens at the highest point on the watershed. This might be the topographical high point, but oftentimes in residential landscapes the highest point on the watershed is the roof of a house. Starting at the highest point, a series of techniques, including contour swales, diversion swales, rain gardens, cisterns and more, make it possible to strategically move the water throughout the site, utilizing and treating water as the precious resource it is.

A great starting point with home-scale landscapes begins with directing downspouts to an aboveground rainwater cistern (or "tank"). From there (or bypassing a cistern altogether), the overflow is directed to rain gardens via underground pipes. As each rain garden fills, they overflow from one to the next and drain into surrounding soils, making the water available to plants near their roots, where they need it.

Recognizing water as a valuable resource means designing landscapes, surfaces, and water systems that help it slow down, soak in, and spread out, ensuring it benefits the environment rather than being lost as runoff. By adopting a few essential techniques, we can increase water retention, minimize waste, and create more resilient ecosystems. In the following chapters, we'll take a closer look at these approaches, including those illustrated opposite.

Organic waste is kept on site in the form of compost to be used for amending soils

Pervious hardscape allows water to pass through and infiltrate the soil

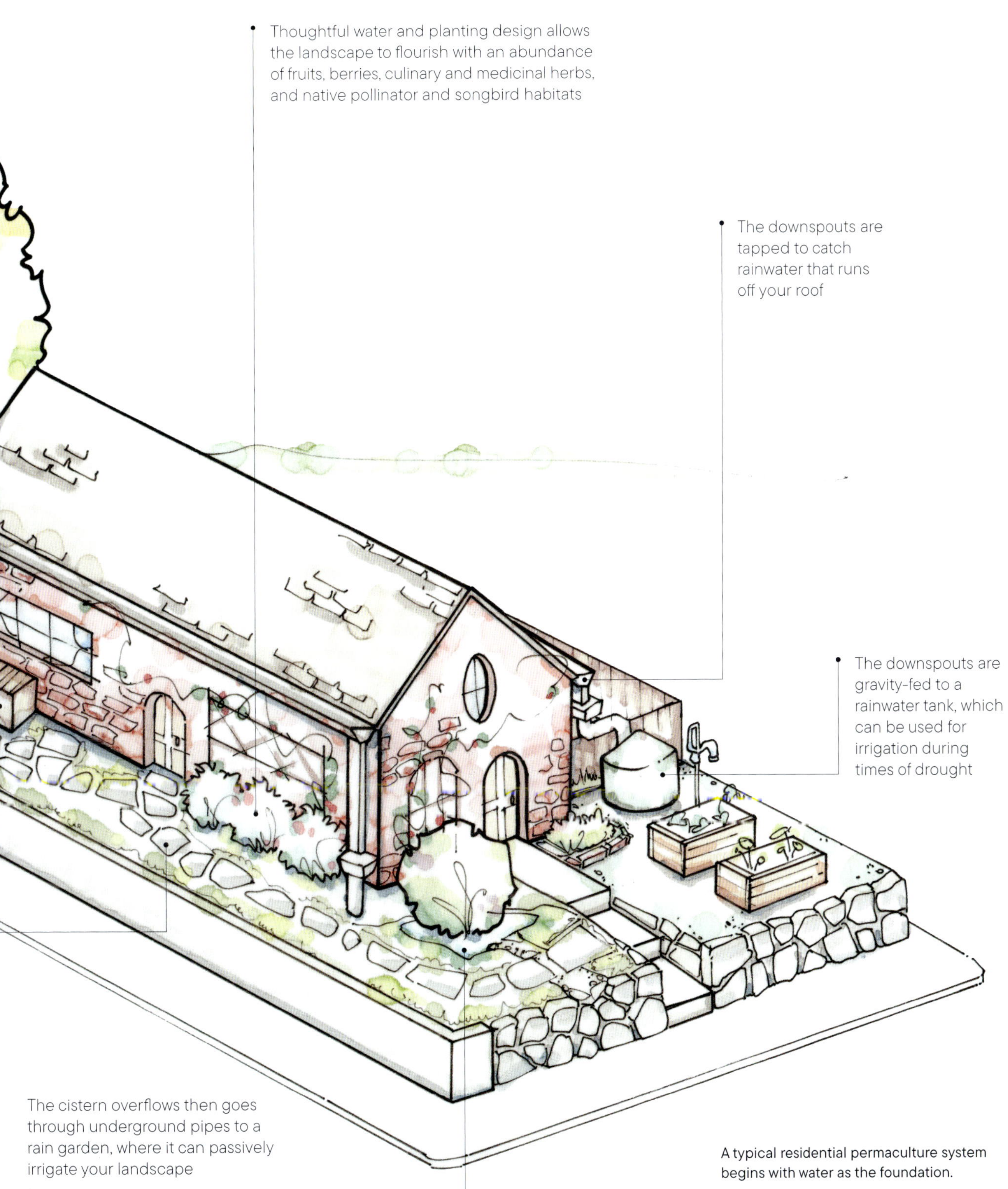

A typical residential permaculture system begins with water as the foundation.

These strategies work together to create healthier, water-efficient landscapes, reducing reliance on external irrigation while strengthening the ecosystem's natural ability to sustain itself.

By applying these strategies, we can work with water rather than against it, ensuring that every drop is used efficiently to support a thriving, regenerative landscape. When managed wisely, water doesn't just sustain life—it amplifies it, supporting the other key pillars of regenerative design.

In section 2, we'll talk about soil health. Healthy soil and water are deeply interconnected. Soil acts as a living sponge, absorbing and storing water while filtering out impurities. The more organic matter and microbial life in the soil, the better it can retain moisture, reducing runoff and the need for irrigation. In turn, well-managed water supports soil health by preventing erosion, replenishing groundwater, and keeping microbial communities alive and active.

In section 3, we'll dive deep into designing plant communities. Plants play a crucial role in water management, using their roots to slow, sink, and spread water into the ground. A diverse plant ecosystem not only prevents soil compaction and erosion but also improves water retention by creating a multilayered, moisture-holding canopy. Deep-rooted plants help access groundwater, while shallow-rooted species stabilize the soil, preventing excess runoff.

Water isn't just a passive element in the landscape—it's an active force that connects and strengthens every other component. By understanding and designing for its natural behaviors, we create landscapes that not only sustain life but actively restore and regenerate it.

Now that you understand the importance of working with water rather than against it, the next step is learning how to apply specific techniques to your own landscape. In chapter 3, we'll introduce key strategies such as swales, rain gardens, and rainwater harvesting. These tools allow you to slow runoff, recharge groundwater, and reduce reliance on municipal water—all while creating lush, productive ecosystems in your yard.

Above Low-volume microemitters work well with rain-fed irrigation systems. They're very efficient, as long as you water late in the evening (provided the water doesn't stay on the leaves) and the water has a chance to sink in before evaporating. Alternatively, subsurface drip irrigation uses a little more water, but is less likely to evaporate because the drip lines are beneath the mulch layer.

Opposite Working with water as a resource makes your entire landscape more resilient in the face of drought and flood. In this photo, the rainwater cistern captures roof water and overflows to rain gardens downhill. The irrigation system of this garden is fed by the cistern, while a heavy layer of mulch and ground cover ensures that the moisture stays in the soil.

A subtle contoured swale along this client's front fence line takes roof runoff and spreads it along the property line, passively irrigating a planting bed filled with native and useful plants such as Culver's root (*Veronicastrum virginicum*), Cardinal flower (*Lobelia cardinalis*), and little bluestem (*Schizachyrium scoparium*).

CHAPTER 3

WATER TECHNIQUES

In this chapter, we dive into the world of earthworks—the art of shaping the land to work with water, not against it. We'll break down three essential techniques, when to use them, and how to assess your site by calculating slope and reading contour lines. Beyond earthworks, we'll explore rainwater harvesting, guiding you through when it makes sense, what factors to consider, and how to put it into practice effectively. By the end, you'll have a deeper understanding of how to move, hold, and use water wisely to create a resilient, regenerative landscape.

What Is an Earthwork?

An earthwork is any intentional modification of the land's surface to guide, capture, or store water. These techniques harness the natural movement of water—shaping the land to slow it down, direct its flow, or hold it in place for future use. By working with gravity and the topography of a site, earthworks help recharge groundwater, prevent erosion, mitigate flooding, and create resilient, self-sustaining landscapes.

Earthworks generally serve one or more of the following purposes:

- Slowing water down to allow it to infiltrate into the soil, preventing runoff and erosion while increasing groundwater recharge.
- Moving water from one area to another, redirecting excess flow away from problem areas or channeling it toward plants and storage systems.
- Retaining water in the landscape by creating features such as ponds, wetlands, or cisterns that store water for later use.

To accomplish these goals, we rely on three core techniques:

1. ***Swales***
   Shallow ditches designed to slow and infiltrate water into the soil.

2. ***Rain gardens***
   Depressions planted with water-loving vegetation that absorb and filter runoff.

3. ***Rainwater harvesting cisterns***
   Above or below ground storage systems that capture and store rainwater.

OBSERVE AND INTERACT: MANAGE WATER ON YOUR SITE

As we've discussed, water is one of the most important elements in any landscape. Managing it well can prevent flooding, reduce erosion, and ensure your plants thrive. The key to good water management is observation. Before making changes, take time to watch how water moves through your space.

Step 1: Observe Water Movement

Evaluate the area. Walk your site during and after rain to see where water collects and where it drains quickly.

Look for pooling areas. These can indicate compacted soil, poor drainage, or low spots that may be ideal for a rain garden.

Check for erosion on slopes, exposed roots, or washouts. These areas need deep-rooted plants or swales to slow water down.

Identify where your roof runoff goes. Downspouts emptying onto pavement or near your foundation can lead to wasted water and moisture issues in your home.

Step 2: Make Adjustments

Slow runoff on slopes. Use swales or deep-rooted vegetation to hold soil and encourage infiltration.

Redirect roof runoff. Extend downspouts into garden beds, rain barrels, or rain gardens to make use of excess water.

Fix pooling areas. Improve drainage by loosening compacted soil, adding organic matter, or planting water-loving species.

Retain water in dry areas. If certain spots dry out too quickly, add mulch, compost, or shade plants to help keep moisture in the soil.

By carefully observing and making simple changes, you can create a landscape that uses water efficiently while preventing damage and waste.

Although cisterns do not always involve reshaping the land, they share the same overarching goal as earthworks: keeping water on-site and using it efficiently.

Before diving into these techniques, we'll first explore some foundational skills that are essential for designing effective earthworks—such as calculating slope and finding contour lines. These skills will allow you to work with the natural flow of water, ensuring that every drop is captured and directed in a way that supports a thriving, regenerative landscape.

CALCULATING SLOPE

Understanding slope is fundamental when designing earthworks, water management systems, and land grading strategies. Slope determines how water moves across the land—whether it infiltrates, flows, or erodes the soil. A site with a gentle slope can effectively absorb and retain water, while a steep slope may require erosion control measures to slow water down. By accurately calculating slope, we can design swales, terraces, retention basins, and other earthworks that work with gravity rather than against it.

Opposite Earthworks can be combined to marry both form and function. In this garden, rain gardens form the base of the planting beds, while permeable pathways drain and replenish soil moisture. This once-dank backyard that had lots of pooling water is now able to manage the water as a resource using a combination of the techniques we'll look at in this chapter.

Above, left A contoured swale along this hillside lets driveway runoff flow into a planting bed, passively irrigating fruit trees and native pollinator plants.

Above, middle A rain garden creates an engineered low spot, so the landscape has somewhere to drain to. Before the rain garden, the runoff from the street above would occasionally flood the house.

Above, right Two small "doorway" tanks are hidden in the side yard of this house. This inverted siphon system takes problematic runoff and sends it uphill to the front yard into a series of rain gardens using only gravity. We will discuss this type of system in depth later in this chapter.

Slope is typically expressed in three ways:

1. ***Rise/Run***
   This is the basic measurement of vertical change (rise) over a horizontal distance (run). It's commonly used in terracing, retaining walls, and drainage calculations to determine how much grading is needed to stabilize an area.

2. ***Ratio (Run:Rise)***
   Often used in civil engineering and erosion control, ratios like 3:1 or 2:1 describe how many feet the land extends horizontally for every one foot of elevation change. A gentle slope might have a 4:1 ratio, while a steep slope could be 1:1 or even greater.

3. ***Percentage (%)***
   This is calculated using the formula
   Slope = (rise ÷ run) × 100

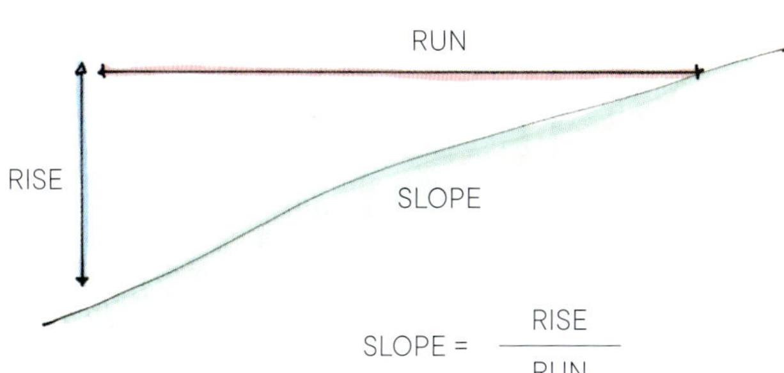

Left Slope is an important calculation to consider when planning water systems and earthworks. The most common way to find this measurement is calculating rise over run.

For example, if the land rises 5 feet (1.5 m) over a 100-foot (30 m) horizontal distance, the slope is 5 percent (5 ÷ 100 × 100). This method is commonly used in stormwater management, grading plans, and accessibility standards.

Why Slope Matters in Earthworks Design

A site's slope dictates water movement, influencing where and how we implement earthworks. For example:

- A **gentle slope** (1 to 5 percent) is ideal for contour swales and infiltration techniques, as water moves slowly enough to be absorbed by the soil.
- A **moderate slope** (5 to 15 percent) requires more structured interventions, such as terracing, check dams, or reinforced swales, to slow and spread water effectively.
- A **steep slope** (15 percent and higher) presents erosion risks, requiring contour planting, terraces, or deep-rooted vegetation to stabilize the land and prevent runoff.

Best Practices for Measuring Slope

When calculating slope, it's essential to keep units consistent—if the rise is measured in inches, the run should also be in inches; if measured in feet, the run should be in feet. Tools like a laser level, A-frame level, or a simple string line and level can help accurately measure slope in the field. With a solid understanding of slope, we can design earthworks that guide water where we want it to go, reducing erosion, increasing infiltration, and creating a more resilient, water-wise landscape.

CONTOUR LINES AND THEIR ROLE IN EARTHWORKS

Contour lines are essential for water management, soil conservation, and regenerative landscape design. A contour line represents a series of points that share the same elevation, forming a level path across the landscape. Because they run perpendicular to the slope and the natural flow of water, contour lines help us understand how water moves through a site and where it can be slowed, redirected, or stored effectively. In practical terms, contour lines are the key to designing swales, terraces, keyline systems, and other earthworks that manage water efficiently and prevent erosion.

For example, when designing a contour swale—a shallow, level trench that captures and disperses water—it is essential that the swale follows a contour line. This ensures that water is evenly distributed across the landscape rather than flowing downhill and eroding the soil. Similarly, terraces are built along contour lines to create flat planting surfaces on sloped land, reducing runoff and increasing water retention.

Finding Contour Lines in the Landscape

If you have access to Geographic Information System (GIS) mapping, topographic surveys, or contour maps, you may already have the data needed to identify contour lines. However, these resources can be expensive and may not be available for smaller residential or community projects. Fortunately, you don't need high-tech equipment to locate contour lines on your site—you can do it using simple tools like an A-frame level or a water level, both of which allow you to identify and mark level lines on the ground.

Opposite The undulating lines of trees in this design represent an orchard laid on-contour, capturing overland water into a series of swales that irrigate the trees during heavy rains.

Right, top Place the A-frame on the ground and adjust until the plumb line indicates level.

Right, middle Mark the ground at both legs. The two points are level with one another.

Right, bottom Pivot on the second point to ensure that any discrepancies are corrected over the distance of your contour line.

Using an A-Frame Level to Mark Contour Lines

An A-frame level is a simple, effective tool made from three sturdy sticks, a string, and a weighted plumb bob or a carpenter's level (see Project: How to Make an A-Frame Level on page 48). This tool allows you to mark precise level points along the land, even on rugged or uneven terrain. By walking the landscape and marking where the A-frame indicates a level point, you can create a continuous contour line, which serves as a guide for earthworks like swales, terraces, or planting beds.

Because it requires no electricity or expensive equipment, the A-frame level is a low-cost, highly accessible tool for regenerative design. It empowers land stewards, farmers, and homeowners to map and shape their land effectively, even in remote or undeveloped areas.

Once you have your A-frame level, you can use it to find the contour lines on your site. Here's how:

- **Mark contour points:** Place the legs of the A-frame on the ground where you want to find a contour. Adjust one leg until the plumb line or bubble indicates the frame is level. Mark the ground at the two points where the legs touch using marking paint or landscape flags.

- **Repeat the process:** Move one leg to the previously marked point and repeat to trace the contour line across your landscape. I recommend pivoting the level each time to adjust for inaccurate calibration. That way, if your level is off by one inch (2.5 cm), for example, it will correct for the one-inch (2.5 cm) discrepancy at the next point, so over the length of your contour line you have an accurate reading.

The Power of Working with Contour

By understanding and utilizing contour lines, we can design landscapes that slow water down, increase infiltration, and prevent erosion. Instead of allowing water to rush downhill, causing soil loss and nutrient depletion, contour-based earthworks help spread water evenly across the land, allowing it to sink in and nourish the soil.

HOW TO MAKE AN A-FRAME LEVEL

Making an A-frame level is straightforward and requires only basic tools and materials. Here's how to create one:

Lay the poles on the ground and add a crossbar, making an "A" shape.

HOW TO BUILD AN A-FRAME LEVEL

1. **Assemble the A-frame.**
   - Lay the two long poles on the ground to form an *A* shape.
   - Attach the shorter pole horizontally between the two longer poles, about 1 to 2 feet (30 to 61 cm) from the top, to serve as the crossbar.
   - Secure the joints with screws, nails, or by tying them tightly with cord.

Materials Needed

→ Two sturdy wooden poles or sticks (4 to 6 feet [122 to 183 cm] long) for the legs
→ One shorter wooden pole or stick (2 to 3 feet [61 to 91 cm] long) for the crossbar
→ A piece of strong string (approximately 3 feet [91 cm] long)
→ A weight (e.g., a small rock or metal washer) for the plumb line
→ Screws, nails, or strong cord to secure the frame
→ A drill or hammer if you're using screws or nails
→ A level

Add the plumb line.

2. **Add the plumb line.**
   - Drill a small hole or attach a hook at the top of the *A* where the two long poles meet.
   - Tie one end of the string securely at this point and attach the weight to the other end to create a plumb line.

Find level ground and mark the point where your plumb line touches the crossbar.

3. **Calibrate the A-frame.**
    - Stand the A-frame on level ground.
    - Allow the plumb line to hang freely and mark the crossbar where the string aligns when the frame is level. This will be your calibration point.

Find another known level location and double check your calibration mark.

4. **Test the A-frame.**
    - Move the A-frame to another location with known level ground and check that the plumb line aligns with your calibration mark. Adjust if needed.

Once built, your A-frame level is ready to use for finding contour lines and leveling terrain!

Swales: Understanding Contour and Diversion Swales

Often in permaculture literature, the swale is touted as the end-all-be-all of permaculture techniques. *Are you even doing permaculture if you don't have a swale?* Like most things, it depends on your goal. And usually, when people are using the word "swale" in permaculture, they're referring to a specific type of swale: a contour swale. But "swale" really just refers to any ditch with a berm, whether it's on-contour or not.

While contour swales are extremely effective at slowing water down and spreading it laterally along a slope, the answer is not always to introduce more water to your landscape. Sometimes your land is saturated, or you experience frequent flooding, and the better strategy for your landscape is to convey the water elsewhere. This is where diversion swales come into play.

In this section, you'll learn the key differences between contour swales and diversion swales, helping you determine which technique is best suited for your site.

CONTOUR SWALES

Contour swales are earthworks built along contour lines to slow water movement and encourage infiltration. They consist of a shallow ditch on the uphill side and a berm on the downhill side, designed to catch and hold water where it naturally flows. Contour swales are often praised as one of the best tools for managing water, but they aren't a one-size-fits-all solution. Understanding your goal—whether it's erosion control, water retention, or groundwater recharge—will help determine whether a contour swale is the right approach for your site.

On a topographic map, contour lines represent points that share the same elevation, illustrating variations in the terrain. Similarly, each point along a contour swale shares the same elevation. Swales are perpendicular to the slope of the land.

To manage overflow, swales can be plumbed together or include a reinforced spillway—a carefully crafted low point in the berm that directs excess water in a controlled way.

The size and placement of swales are determined by the unique conditions of a site. To ensure even water distribution, using smaller swales at more frequent intervals is often the most effective approach. However, when space is limited and larger volumes of water need to be managed, a single, deeper swale may be the more suitable option.

A contour swale's main purpose is to slow down surface water flow and allow it to absorb into the soil. As water moves downhill, it gradually fills the swale along its length. Once the swale reaches capacity, the excess water either spills over the berm, exits through a designated spillway, or is channeled through a pipe for redirection further downslope.

Contour swales are a very effective tool for mitigating erosion and spreading water evenly across a site. They are also an excellent tool for passively irrigating a large planting area, such as an orchard.

When to Use Contour Swales

Contour swales are most effective in dry, open landscapes where slowing and infiltrating water is a priority. They work well on gentle to moderate slopes by spreading water evenly across the landscape, allowing it to soak into the soil rather than running off. Swales are most effective on slopes with a gradient of

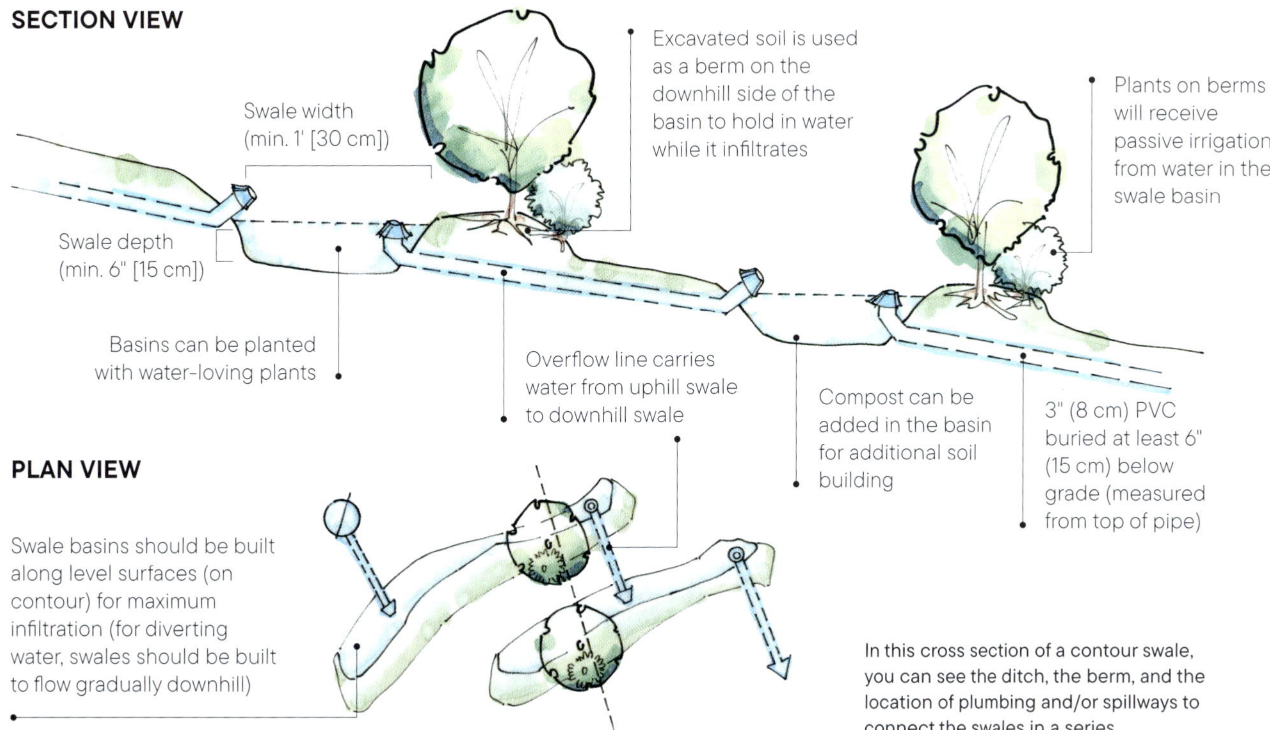

In this cross section of a contour swale, you can see the ditch, the berm, and the location of plumbing and/or spillways to connect the swales in a series.

3:1 or gentler, as this helps prevent the berm from becoming too steep, which could cause it to erode. Swales are particularly useful when placed higher in a watershed, where they can capture and slow rainfall before it gathers momentum, preventing flooding and erosion further downhill. By intercepting water early, they help recharge groundwater and keep moisture available for plants in drier times.

When NOT to Use Contour Swales

While swales are a powerful tool, they aren't always the right solution. Avoid using them in areas where excess water is already a problem, as they are designed to increase infiltration, not remove water. If the goal is to drain water from a site rather than retain it, other solutions—such as drainage channels or rain gardens—may be more appropriate.

Contour swales are also not ideal in heavily forested areas, where disturbing tree roots can damage soil structure and disrupt established ecosystems. In such environments, a less invasive alternative is to lay brush along contour lines, which slows water without major excavation.

Contour swales may not be as effective on sites with multiple features like fences, hardscaping, or other obstacles, especially if there isn't enough uninterrupted space for their construction. Since they naturally follow the land's curves, it's important to thoughtfully plan access points and pathways to ensure smooth navigation throughout the area.

On steep slopes exceeding 15 percent, swales can become unstable, as berms are more likely to fail under heavy water flow. Instead, consider using brush laid on-contour, small check dams, or deep-rooted vegetation to stabilize the soil and slow water naturally without the risk of erosion or collapse.

Opposite Topographic maps show contour lines, illustrating variations in the terrain. Each contour line represents a series of points at the same elevation.

HOW TO BUILD A CONTOUR SWALE: A STEP-BY-STEP GUIDE

Creating a contour swale on your land involves a series of interconnected steps, each crucial to the success and effectiveness of your earthworks.

Using an A-frame level, mark a contour line where you want to put your swale. If you are intending to mark multiple swales, you may need to play with the starting point a few times to make sure the layout works in your landscape. Using flags instead of paint to mark makes changing the layout a little easier. When you're ready, cut along the line and pile the dirt on the downhill side of the line to begin forming the berm.

Cut your first line along the contour line you just marked. Begin excavating uphill of the line and piling the dirt on the downhill side of the line. Start breaking up the soil clods, especially if you're working with heavy clay, and sculpting the berm. Check your level as you go. I like to use a board with a level on top so I can get a visual of where the gaps in the berm are and correct them as I go. When you finish and you have ensured your berm is level, compact the soil and choose how you will stabilize the dirt to avoid a washout.

SITE ASSESSMENT AND PLANNING

- **Understanding Your Landscape.** Begin by carefully observing your land's topography, soil composition, and existing vegetation. Identify areas with slopes and potential water flow pathways.
- **Marking Contour Lines.** Using your A-frame level, mark contour lines along the slope. These lines will guide the placement of your swale.
- **Determining Swale Location and Size.** Based on your site assessment, choose the optimal location for your swale. Consider factors such as total water catchment area, soil type, and desired function of the swale (water retention, erosion control, etc.). Determine the appropriate size and depth of the swale based on these factors.

MARKING AND EXCAVATION

- **Marking the Swale.** Using stakes and string, mark the downhill line of the swale along the contour.
- **Excavation.** Carefully excavate the swale trench along the marked outline. The excavated soil can be used to build a berm on the downhill side of the line. As you excavate, don't worry about making the bottom of the ditch level. You will do the leveling on the berm instead. Remember that water is constantly seeking its own level, so even though the swale may fill unevenly if the bottom of the ditch isn't level, that doesn't affect its function because the berm and overflow will set the water level.

BERM CONSTRUCTION

- **Building the Berm.** Use the excavated soil to construct a berm on the downhill side of the swale. The berm helps to slow down water flow and retain moisture. Compact the dirt as you go, making a gentle slope to the berm as it ties back into the grade downhill.
- **Shaping and Stabilizing the Berm.** Shape the berm to an appropriate height and slope. Stabilize the berm with vegetation, rocks, or other natural materials.
- **Leveling the Berm.** Once you have your berm "roughed in," take a long board, such as a 2-inch by 4-inch by 6-foot (5 cm by 10 cm by 1.8 m) piece of lumber, and place a hand level on top. Crouch down to see where your berm has gaps between the top of the berm and the bottom of the board. You also want to confirm that your berm is level from end to end. This might involve dropping one side, or building up one side, depending on how even your excavation was.

SWALE ENHANCEMENT AND PLANTING

- **Incorporating Organic Matter.** Add compost, mulch, or other organic matter to the swale to improve soil fertility and water retention. You can use an engineered soil mix in the swale, like you would in a rain garden, to create a permeable planting mix. Keep in mind any material you add into the swale will reduce the volume of water that it can hold. I also recommend keeping at least 3 inches (8 cm) of "ponding depth," meaning the space between the level of the overflow and the top of the organic material.
- **Planting Vegetation.** Plant a diverse range of vegetation in and around the swale. Choose plants that are suited to the local climate and soil conditions. Consider using a mix of deep-rooted plants, ground covers, and nitrogen-fixing plants. I generally recommend planting trees and shrubs on the downhill side of swales, rather than on the berms directly, so as not to compromise the integrity of the berm. Reserve the berms and basins for bunching grasses, ground covers, and perennials.

MAINTENANCE AND MONITORING

- **Regular Inspection.** Regularly inspect the swale for signs of erosion or damage. Sometimes after large rain events, especially while plants are establishing, you might see that the berm has "blown out." That's okay! Just add soil and relevel as needed.
- **Keep your overflow pipe and/or spillway clear of debris.** The level berm functions as a secondary overflow if needed, but until vegetation is fully established, the berm will be more likely to blow out if your primary overflow is clogged.
- **Vegetation Management.** Maintain healthy vegetation in and around the swale.
- **Sediment Removal.** Remove any sediment that accumulates in the swale. If you're seeing a lot of sedimentation, that's an opportunity to look uphill and see where the erosion is coming from so you can stabilize it.
- **Adjustments and Improvements.** Make any necessary adjustments or improvements to the swale based on your observations and monitoring.

DIVERSION SWALES

Water is a valuable resource, and the goal is to retain as much of it on-site as possible, sometimes up to five inches (13 cm) of rain at a time. However, there comes a point where too much water in the wrong place can become a problem. When excess water pools in areas where it can cause damage or flooding, a diversion swale can be an effective solution.

The common approach to managing runoff is to deal with it at the lowest point—often through French drains, sump pumps, or other drainage systems. But let's rethink that. Time and again, I've worked with homeowners who have serious drainage issues, only to find that previous solutions simply redirected water to the lowest spot, where gravity no longer works in their favor. Too often, that low point ends up being the home's foundation or crawl space, leading to costly flooding and repairs should the intervention fail.

A diversion swale takes a different approach by rerouting water before it becomes a problem. Instead of allowing runoff to rush straight downhill and collect in one area, diversion swales direct the flow to a more suitable location, allowing some of the water to slow down and infiltrate into the soil along the way. Structurally, they resemble contour swales, featuring a ditch with a berm on the downslope side. However, unlike contour swales, diversion swales do not follow the contour line. Instead, they can be positioned slightly off-contour (around a 1 percent slope) or even perpendicular to the land's natural grade to efficiently channel water to a designated area.

Diversion swales can be lined with stone or planted with deep-rooted, resilient vegetation that thrives in both drought and flood conditions. Plants such as muhly grass (*Muhlenbergia capillaris*) are excellent choices for stabilizing the swale and enhancing absorption. Once the water has traveled as far as needed, it can either be piped away or allowed to percolate into a rain garden.

Types of Diversion Swales

Diversion swales are designed to redirect water flow away from sensitive areas such as foundations, roads, or eroding slopes, rather than dispersing it into the soil like contour swales. They are typically built slightly off-contour (1 percent to 5 percent slope) to encourage water movement while still preventing excessive runoff and erosion. Two common types of diversion swales are bioswales and dry creek beds, each serving a distinct function in water management.

Bioswales: Slow, Filter, and Move Water

Bioswales are vegetated swales that move water gradually while filtering out pollutants and sediments. They are commonly used in stormwater management, particularly along roads, driveways, and urban areas where runoff may contain oils, heavy metals, or other contaminants. By integrating deep-rooted plants, mulch, and permeable soils, bioswales slow water flow, allowing sediment to settle and nutrients to be absorbed before the water reaches a drainage system, wetland, or water body.

Bioswales are especially valuable in areas where water quality is a concern, as they function like natural filtration systems. They also add ecological value by providing habitat for pollinators and beneficial soil organisms.

Opposite A vegetated diversion swale, or bioswale, is planted densely while directing water somewhere useful in the landscape. This one flows along a retaining wall and terminates at a small basin that then pipes the water to the front yard into a series of rain gardens.

Right A dry creek bed slows water while directing it downhill and away from the home's foundation.

Dry Creek Beds: Guiding Water with Natural Aesthetics

Dry creek beds are stone-lined swales designed to channel water from one point to another, mimicking the look of a natural creek. Unlike bioswales, which prioritize filtration and infiltration, dry creek beds focus on efficiently guiding water while reducing erosion. They are useful in areas where heavy runoff needs to be redirected, such as the base of hills, alongside structures, or near downspouts.

- On gentle slopes (5 percent or less), a dry creek bed passively directs water while allowing some infiltration.
- On steeper slopes, infiltration basins beneath the creek bed act as underground speed bumps, slowing water down before it continues downslope. This prevents excessive runoff and minimizes erosion risks.

Design Considerations for Diversion Swales

A well-designed dry creek bed should match the scale of the watershed to handle the volume of water it receives. Undersized creek beds can overflow or erode, causing unintended damage to the landscape. A good rule of thumb is to err on the side of oversizing, especially in regions prone to heavy storms. Additionally, steeper slopes require heavier, larger stones to resist displacement during fast-moving flows.

By understanding the different types of diversion swales and their applications, we can design water management systems that balance function, aesthetics, and ecological benefit, helping to protect soil, manage runoff, and improve water quality across various landscapes.

When to Use Diversion Swales

Diversion swales are most effective when the goal is to redirect water from one area to another. They move water efficiently while minimizing erosion and managing runoff. However, because they transport water rather than absorbing it, it's essential to understand the volume of water you're dealing with and have a clear plan for where it will go and how it will be managed at its final destination. This is where rain gardens come into play, providing a place for redirected water to settle and absorb into the landscape.

Diversion swales are a great solution for sites experiencing:

- **Pooling or flooding in low-lying areas.** By intercepting runoff before it accumulates in depressions or near structures, diversion swales help mitigate flood risk and waterlogging.
- **Erosion-prone slopes.** On sloped landscapes, unmanaged water can gain momentum and strip away soil. A well-placed diversion swale can help

stabilize these areas by controlling flow and directing water to a safer location.

- **Redirecting water to storage or infiltration areas.** If you need to guide water away from foundations, driveways, or compacted zones, diversion swales can direct it toward rain gardens, ponds, or permeable surfaces where it can be absorbed and utilized.

When NOT to Use Diversion Swales

Although effective for moving water, diversion swales are not always the right tool, particularly when water infiltration is the primary goal.

- **If you want the water to penetrate directly into the soil.** If groundwater recharge or passive irrigation is the goal, contour swales or infiltration basins may be more appropriate.
- **In heavily forested areas.** In dense woodlands, excavation for a swale can disrupt tree roots, compact soil, and alter natural drainage patterns. Instead, consider using brush laid slightly off-contour to slow and spread water without disturbing the existing ecosystem.

By carefully considering the purpose and placement of a diversion swale, you can improve drainage, protect soil, and manage water effectively, ensuring that excess runoff is handled in a way that benefits the surrounding landscape rather than causing unintended damage.

Understanding the differences between contour and diversion swales allows you to manage water efficiently across your landscape. The key is to observe water patterns on your site and think creatively about how to guide it. Working with water may seem simple, but it often challenges our instincts—so take your time, experiment, and enjoy the process of dancing water through your land.

WHAT IS ENGINEERED SOIL?

You'll hear me reference "engineered soil" frequently, especially when I'm talking about water techniques. Engineered soil is a man-made blend of various components designed to optimize plant growth, regulate water movement, and support environmental sustainability. Unlike naturally formed soil, which evolves through organic decomposition and geological activity, engineered soil is purposefully created by mixing materials such as sand, clay, compost, and biochar to achieve specific qualities.

This type of soil is widely applied in urban settings, ecological restoration projects, and agriculture to enhance soil structure, improve moisture retention, and facilitate drainage. In cities, for example, it is commonly found in green roofs and rain gardens, where it helps manage stormwater while providing a fertile base for plants. Additionally, engineered soil can be tailored to reduce compaction, encourage beneficial microbial life, and aid in carbon sequestration, making it an essential tool in sustainable landscape design and climate resilience efforts.

Engineered soil can have different ratios of sand:compost or sand:topsoil, depending on the application. I have found that the ratio you use depends on what you are planting and how drought tolerant it is. The more sand you have in your mixture, the less water the soil will hold over time. I've experimented with a lot of recommended mixtures and have landed on one that I think works best, balancing plant health while allowing proper water infiltration and retention: 50 percent sand and 50 percent compost.

Where Do You Buy Engineered Soil?

For quantities you would use at a residential scale, the best option is to ask a local bulk landscaping materials supplier to premix your blend with a 1:1 ratio of sand:compost.

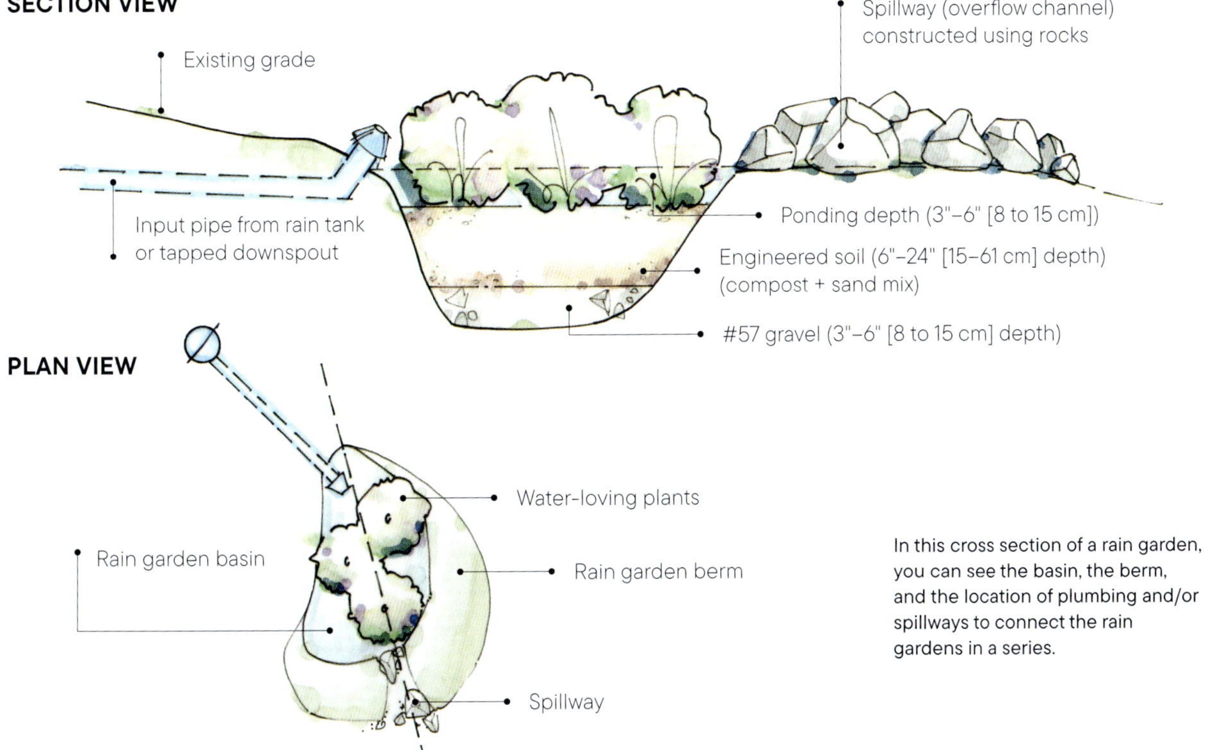

In this cross section of a rain garden, you can see the basin, the berm, and the location of plumbing and/or spillways to connect the rain gardens in a series.

Rain Gardens

As I mentioned earlier, the techniques in this chapter can be combined to fit your site's specific needs and constraints. Rain gardens are a great way to strategically direct water into the soil in targeted areas. They can collect runoff from the land, such as water flowing through a diversion swale, or capture water from your roof through diverted downspouts. One of the benefits of rain gardens is that they make it easy to track how much water is being absorbed while seamlessly integrating into your landscape without interfering with paths, play areas, or other features.

A rain garden is a carefully designed shallow basin that captures rainwater and helps it absorb into the soil. By harnessing natural processes, rain gardens help manage excess water while supporting biodiversity, soil health, and groundwater recharge. Their structure includes several key elements that work together to ensure efficient water management.

ANATOMY OF A RAIN GARDEN

Unlike a pond, which retains water for extended periods, a rain garden is a basin with a berm that temporarily holds water during heavy rains, allowing it to sink into the surrounding soil. Typically irregular in shape, rain gardens are engineered to drain completely within about twelve hours after a rain event, depending on the soil type and saturation levels.

Rain gardens are commonly incorporated into green infrastructure projects, where they are referred to as "bioretention areas." Unlike ponds, they don't retain water permanently—instead, they capture runoff and drain completely after a rainfall. While there are different ways to build a rain garden, most designs include a gravel sub-base, an engineered soil mix (typically 50 percent sand and 50 percent planting soil), and a designated ponding depth to temporarily hold water before it soaks into the ground. Each of these strata has a certain percentage of void space, and adjusting the depths changes how much water the rain garden can hold.

Feedline (Input/Inflow)

The feedline directs water into the rain garden, typically using a three-inch (8 cm) Schedule 40 PVC pipe. This pipe channels overflow from cisterns, downspouts, or other runoff sources, ensuring water enters the basin in a controlled manner. Proper placement of the feedline prevents erosion at the entry point and evenly distributes water into the rain garden.

Overflow Line (Standpipe)

Since rain gardens are designed to fill during heavy rain events, an overflow line is essential to prevent flooding and structural damage. This feature consists of a capped PVC standpipe with an atrium grate, which allows excess water to flow out and be redirected to another rain garden or dispersed into the landscape.

When multiple rain gardens are installed in sequence, they function like ice cube trays, filling one after the other as each basin reaches capacity. If a rain garden is the final stop in a system, the berm itself acts as the overflow mechanism, allowing excess water to gently spill out into the surrounding landscape.

Berm

The berm is a slightly raised mound of compacted soil that surrounds the rain garden, creating a barrier that helps contain water. It is formed from excavated soil and gradually blends into the existing landscape, preventing blowouts while directing and slowing overflow. If the standpipe becomes blocked, the berm serves as a secondary overflow structure, allowing water to spill over in a controlled manner rather than eroding the garden.

Ponding Depth

Ponding depth refers to the amount of water the rain garden can hold before it reaches the overflow point. This depth excludes any space occupied by gravel, soil, or other materials—it represents the actual volume of stormwater that the garden is designed to manage. The deeper the basin, the more water it can hold and slowly release into the soil, reducing runoff and improving groundwater recharge.

Basin

The basin is the excavated area where water collects, and it consists of two distinct planting zones that accommodate varying moisture levels:

- **Wet Zone:** The lowest part of the rain garden, which is temporarily submerged during rain events. Plants here must tolerate both standing water and dry periods.
- **Mesic Zone:** Surrounding the wet zone, this area experiences periodic saturation but does not get fully submerged, making it ideal for plants adapted to fluctuating moisture levels.

Through the careful selection of native, deep-rooted plants, the basin enhances water infiltration, prevents erosion, and creates habitat for pollinators and beneficial organisms.

Infill Material

Rain gardens are designed to maximize water absorption and filtration, using a combination of materials with different water-holding capacities:

- **Gravel:** Occupies about 75 percent void space, providing excellent drainage and preventing soil compaction.
- **Engineered Soil:** Contains 25 percent void space, ensuring a balance between drainage and water retention for plant health.
- **Ponding Depth:** Represents 100 percent available space for water storage, allowing rainwater to settle before slowly filtering into the ground.

By incorporating these structural elements, a rain garden can effectively manage stormwater, reduce erosion, improve soil health, and support local biodiversity—all while enhancing the beauty and function of the landscape.

WHEN NOT TO USE A RAIN GARDEN

Like every permaculture technique, rain gardens are not suitable for every location. Proper siting is critical to

You can use stone or plants to hide the piping. Well-crafted rain gardens are barely noticeable in a landscape.

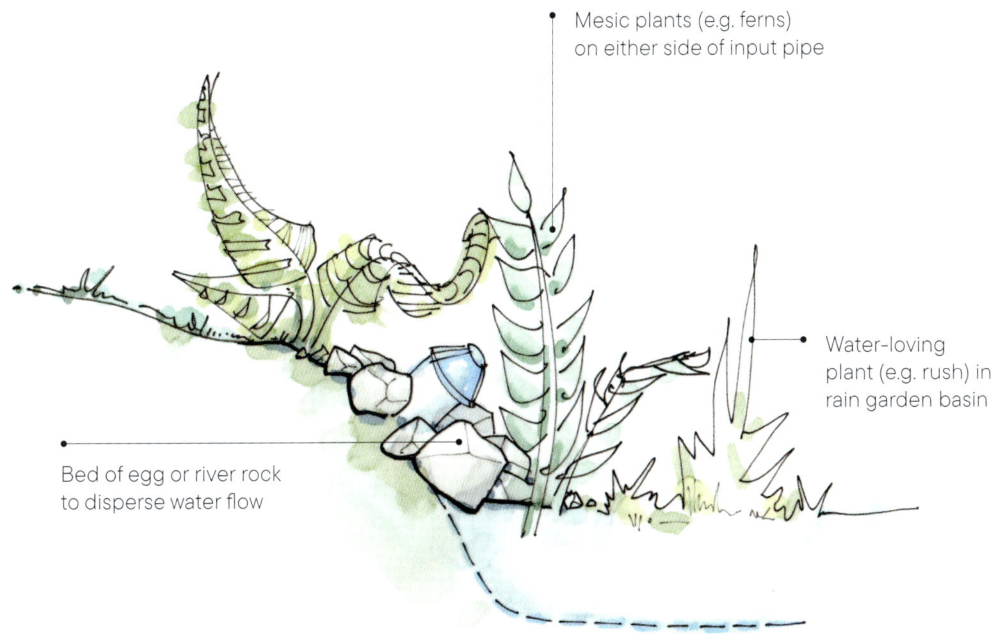

Rain gardens have some areas that stay more moist than other areas. This illustrates the mesic and dry zones of a rain garden, which can guide you as you build your plant palette. Rain gardens are designed to fully drain, so there are no areas that are always wet.

prevent unintended consequences, such as structural damage, root stress, or poor drainage issues. Following are key considerations for when *not* to install a rain garden.

Too Close to a Building Foundation

Because rain gardens are designed to hold and disperse significant volumes of water, placing them too close to a house or structure can lead to foundation damage, basement flooding, or unwanted moisture buildup. Water seepage near a foundation can cause cracking, mold growth, and structural instability over time.

Guideline: Install rain gardens at least fifteen feet (4.6 m) away from any building foundation, increasing the distance for larger rain gardens that hold more water.

Within a Tree's Critical Root Zone

Mature trees play a crucial role in the landscape, stabilizing soil, providing shade, and absorbing water. However, excavation during rain garden construction can disturb the critical root zone, damaging delicate feeder roots and compromising tree health. Trees stressed by excess water or soil disruption may become susceptible to disease, pests, or decline.

Guideline: Avoid placing rain gardens within the critical root zone, which typically extends one to three times the tree's dripline (the outermost edge of its canopy). Instead, choose a site where excavation won't interfere with major root structures.

In Areas with Poorly Draining or Compacted Soil

Rain gardens are meant to allow water to infiltrate soil, but if the soil has high-clay content, severe compaction, or a shallow hardpan layer, water may pool and stagnate instead of absorbing. This can lead to standing water, mosquito breeding, and plant stress.

Guideline: Conduct a percolation test before installing a rain garden. If water drains at less than half an inch (1.3 cm) per hour, consider amending the soil, installing a drainage layer, or choosing an alternative water management strategy.

In Areas Prone to High Groundwater Tables

If the water table is too close to the surface, a rain garden may not drain properly, resulting in persistent waterlogging and plant failure. Instead of dispersing stormwater, the basin may simply fill up like a pond, defeating its purpose as a temporary retention feature.

Guideline: Avoid placing rain gardens in areas where the seasonal high-water table is closer than three feet (91 cm) below the surface. If you're unsure, dig a test pit and look for signs of water saturation before installation.

On Steep Slopes Without Proper Design Adjustments

While rain gardens work well on gentle slopes, installing them on steep inclines (above 10 percent) without modifications can lead to erosion, berm failure, and uneven water distribution. On steeper terrain, water rushes through too quickly, preventing proper infiltration.

Guideline: If placing a rain garden on a slope, consider terracing the landscape, using check dams, or incorporating a series of smaller basins to slow water movement.

Where It May Interfere with Underground Utilities

Excavating a rain garden without checking for underground utilities can result in costly and dangerous damage to sewer, water, gas, or electrical lines.

Guideline: Before digging, always call your local utility location service to mark underground lines and ensure a safe, legal installation.

By carefully evaluating the site and avoiding these common pitfalls, you can ensure that your rain garden functions effectively, enhances the landscape, and supports long-term sustainability without unintended drawbacks.

Real-Life Applications of Rain Gardens

ACCOMMODATING ROOF AND OVERLAND WATER

In one of our residential projects, the front yard downspouts are plumbed together to feed into rain gardens in the front yard. The excavated basins were filled with gravel and engineered soil, surrounded by berms of native soil, and planted with suitable vegetation.

The backyard uses a series of small rain gardens near the corner of the property to slow overground water that was causing a lot of erosion. These basins, although they look small, are 3 to 4 feet (91 to 122 cm) deep and accommodate several thousand gallons of water in a compact area. They capture both overland flow and roof runoff and are planted with shade-tolerant sedges (*Carex* spp.) and ferns. The rain gardens fill and overflow sequentially, demonstrating effective water management of the downspout water from approximately 1,500 sq ft (140 sq m) of roof area. Something else notable about these rain gardens is that there is no uphill berm, which allows them to also receive overland flow as part of an integrated water management approach, capturing both overland flow and roof water.

ACCOMMODATING CISTERN OVERFLOW

At a local community center food forest, rain gardens manage overflow from a rainwater harvesting cistern. They are fed by the overflow from a 1,600-gallon (6,057 L) rainwater cistern that captures 5,000 sq ft (465 sq m) of roof water where one inch (2.5 cm) of rain generates 3,100 gallons (11,735 L) of roof runoff. When sizing earthworks that collect the overflow from a rain cistern, it's important to assume the cistern is full (as it will be in many months of the year when you aren't irrigating), that way your earthworks can accommodate the volume of water they will be receiving from the roof. These rain gardens were crafted using leaf mulch and hardwood mulch as the infill material, rather than gravel and engineered soil. The plants, mostly fruit trees, were planted on the berm and around the basin instead of within the basin.

MITIGATING DRAINAGE ISSUES

In one of our projects, we were called in to correct drainage and erosion issues less than six months after construction. Mud was consistently collecting in their driveway. The corner lot also has great sunlight, and the client wanted to reduce the expanse of lawn to grow fruits, berries, and pollinator habitat.

Far left Downspout filters filter out debris and mosquito larvae on the feedlines from this roof. The PVC pipes run underground to a series of rain gardens that also capture overland water.

Left These woodland rain gardens, out of the critical root zone of overstory trees, capture both roof and overland water. They are densely planted, allowing them to blend into the surrounding woodland understory.

Left, top Several years after the initial installation, the orchard plantings of Illinois everbearing mulberry (*Morus alba × rubra* 'Illinois Everbearing'), fig (*Ficus carica*), pomegranate (*Punica granatum*), apple (*Malus domestica*), and elderberry (*Sambucus nigra*) have thrived with very minimal irrigation, relying almost entirely on the gravity-fed passive irrigation of the cistern overflow.

Left, bottom The rain garden pictured is planted heavily with iris (*Iris* spp.), elephant ear (*Colocasia* sp.), and river oats (*Chasmanthium latifolium*), which can take both flood and drought conditions.

Opposite A 1,000-gallon (3,785 L) cistern captures roof water, directs overflow to rain gardens, and feeds a drip irrigation system with a solar-powered pump.

We tapped the downspouts and gravity-fed them into a cistern, which overflowed to the front yard into a series of rain gardens.

We placed hydrants for rain-fed irrigation from the cisterns. We then diverted water from the driveway via a bioswale, which turned on-contour to passively irrigate fruit trees and infiltrate into the hillside. Slowing the water along the slope created an opportunity to grow many native plants and hydrate the soils.

We also engineered a low spot along the driveway to give the overland water somewhere to collect and infiltrate from, since this client's property was at the very bottom of the watershed of four newly constructed homes.

As you can see, rain gardens often work in tandem with other techniques, such as contour swales and diversion swales, to move, slow, and spread water strategically through a landscape. Rain gardens are a versatile and efficient way to manage water in the landscape. By combining them with other earthworks, you can create a cohesive system that slows, sinks, and spreads water effectively, transforming potential runoff problems into opportunities for soil hydration and plant growth. In the next chapter, we'll explore rainwater harvesting and how to integrate cistern overflow into your landscape.

Rainwater Harvesting

Water is needed in any landscape where plants grow, and rainwater harvesting allows us to capture, store, and reuse it. A well-designed rainwater harvesting system ensures that every drop is put to good use. Harvested rainwater is an incredibly versatile resource that can be used for a variety of purposes, depending on the level of filtration and treatment. By integrating rainwater harvesting into your landscape, you can reduce reliance on municipal water, conserve resources, and create a more self-sufficient system.

WHAT IS A RAINWATER HARVESTING SYSTEM?

A rainwater harvesting system collects and stores rainwater for later use. It can be as simple as a rain barrel or as complex as an in-ground cistern with a pump. Systems can distribute water using gravity or pumps, depending on their design and purpose.

For example, a 75-gallon (284 L) barrel might use gravity to supply water for foot washing, dog watering, or mud kitchen play. On the other hand, a 300-gallon (1,136 L) aboveground cistern with a submersible pump can provide irrigation through a yard hydrant.

COLLECTED RAINWATER USES

Irrigation and Landscape Watering

- One of the most common uses of collected rainwater is watering gardens, orchards, lawns, and native landscapes.
- Drip irrigation or gravity-fed systems can efficiently distribute stored rainwater directly to plant roots.
- Because rainwater is naturally soft (free of chlorine, fluoride, and salts found in municipal water), it promotes healthier soil and plant growth.

Wildlife and Habitat Support

- Collected rainwater provides drinking water for birds, pollinators, and beneficial wildlife.
- Rainwater can be directed into ponds, wetlands, or habitat restoration areas to create microclimates and biodiversity hubs.
- Collected rainwater also supports aquatic plants and fish when stored in properly designed systems.

Household Non-Potable Uses

With minimal filtration, rainwater can be used for many household functions, reducing dependence on treated municipal water:

- Toilet flushing (one of the largest household water consumers).
- Laundry, as rainwater is free of minerals that cause soap buildup.
- Cleaning, car washing, and exterior washing (such as patios, sidewalks, and tools).

Drinking and Cooking (with Proper Treatment)

When properly filtered and treated, rainwater can be made safe for human consumption.

- Advanced filtration (sediment filters, activated carbon, UV, and reverse osmosis) removes contaminants.
- Proper storage in dark, sealed cisterns prevents algae and bacterial growth.

Emergency and Drought Resilience

- A stored supply of rainwater can serve as backup water during droughts, municipal shortages, or emergencies.
- Rainwater can be a reliable water source for rural or off-grid homes that lack access to consistent groundwater or municipal supply.

By designing a thoughtful rainwater harvesting system, you can make the most of this free, abundant resource, increasing sustainability and resilience in your home and landscape.

WHEN IS A RAINWATER HARVESTING SYSTEM USEFUL?

A rainwater harvesting system is a valuable tool for reducing water consumption, increasing landscape resilience, and making use of a free, natural resource. It is particularly useful in situations where non-potable water can replace or supplement municipal, well, or surface water sources.

Consider installing a rainwater harvesting system if:

- You have regular non-potable water needs, such as irrigation, washing, water play, or outdoor cleaning. Rainwater is naturally soft and free of chlorine, making it ideal for watering plants and reducing buildup in cleaning applications.
- You live in an area with sporadic or seasonal rainfall and want to store water for use during dry periods. A well-sized system can provide water security during droughts and prevent over-reliance on municipal supplies.
- You maintain a large vegetable garden, orchard, or lawn and want to reduce dependence on well water or municipal supply while keeping plants healthy.
- You want to reduce stormwater runoff on your property, preventing erosion, flooding, or nutrient loss while putting excess water to beneficial use.

The size of the system should match your water needs:

- Small-scale systems (e.g., rain barrels, typically 50 to 100 gallons [189 to 379 L]) are great for potted plants, occasional hand watering, or outdoor cleaning.
- Medium-sized systems (e.g., tanks holding 300+ gallons [1,136+ L] with a gravity-fed or pump system) are suitable for gardens, lawn care, and moderate irrigation.
- Large systems (e.g., cisterns storing 1,000+ gallons [3,785+ L]) are ideal for extensive irrigation, livestock watering, or integrating rainwater into household non-potable uses like flushing toilets or washing clothes.

WHEN IS A RAINWATER HARVESTING SYSTEM <u>NOT</u> USEFUL?

While rainwater harvesting is a powerful tool, it is not always the best solution. There are certain situations where other water management strategies might be more effective:

- If you have high irrigation needs and reliable well water, it may be more sustainable to focus on groundwater recharge and infiltration rather than installing a large rainwater harvesting system.
- If the catchment area is too small to meet your water needs, such as a small roof with limited runoff potential, the amount of water collected may not justify the cost or effort of installing a storage system.
- If rainfall is minimal and irregular, a small-scale system may not store enough water to be useful for high-demand landscapes.

Ultimately, the effectiveness of a rainwater harvesting system depends on site conditions, water demand, and long-term sustainability goals. By carefully assessing these factors, you can determine whether rainwater harvesting is the right solution or if other water management techniques—such as earthworks, soil building, or efficient irrigation—might serve your landscape better.

ANATOMY OF A RAINWATER HARVESTING SYSTEM

Gutters and Downspouts

Gutters and downspouts are key components of a building's rainwater drainage system, designed to collect and direct rainwater away from the roof and foundation.

- Gutters are horizontal channels attached along the edges of a roof that catch and guide rainwater toward the downspouts.
- Downspouts are vertical pipes that carry the collected water from the gutters down to the ground or into a designated drainage or storage system.

Tanks, Cisterns, and Rain Barrels

Tanks, cisterns, and rain barrels are storage systems used to capture and hold rainwater for later use, whether for irrigation, household needs, or emergency water supply. While they serve the same basic function, they vary in size, material, and capacity to suit different applications.

Rain Barrels

- Small-scale storage (typically 50 to 100 gallons [189 to 379 L]).
- Ideal for gardens, potted plants, and hand watering.
- Often placed at the base of a downspout, collecting water directly from gutters.
- Typically made of plastic, metal, or repurposed food-grade barrels.
- May include spigots, overflow valves, and hose attachments for easy use.

Cisterns

- Medium-to-large storage systems (ranging from 300 gallons [1,136 L] to several thousand gallons).
- Used for landscape irrigation, livestock watering, and even household non-potable uses (e.g., flushing toilets, laundry).
- Can be above ground or underground, depending on space and climate considerations.
- Made from concrete, metal, fiberglass, or heavy-duty plastic.
- Often equipped with filtration, pumps, and overflow management systems.

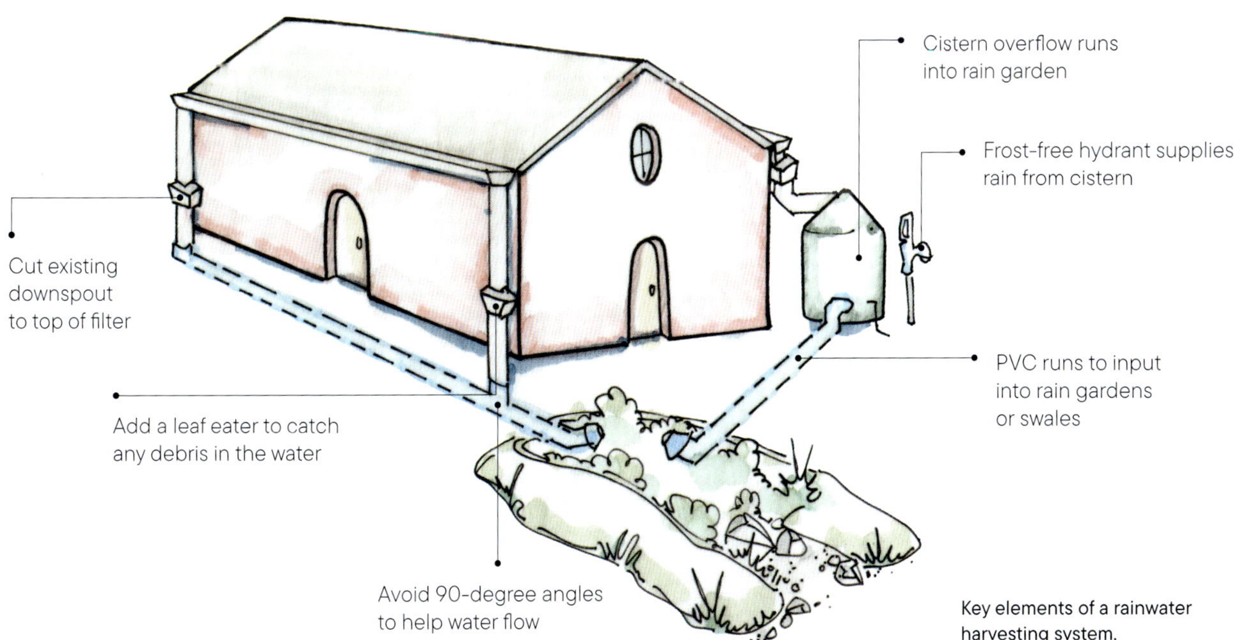

Key elements of a rainwater harvesting system.

Tanks

- A broad term encompassing any large storage vessel for rainwater collection.
- Includes both cisterns and high-capacity storage tanks (10,000+ gallons [37,854+ L]).
- Used in large-scale irrigation, potable water supply (with proper filtration), fire suppression, and commercial applications.
- Can be integrated into off-grid water systems or used as backup water reserves in drought-prone areas.

Downspout Filters

Downspout filters, commonly known as leaf eaters, are devices installed on downspouts to screen out leaves, debris, and insects before rainwater enters a storage system like a rain barrel, cistern, or tank. They are a crucial first-stage filtration component in rainwater harvesting, ensuring cleaner water collection and reducing maintenance needs.

How They Work

- Mounted on the downspout where water exits the gutter.
- Use a mesh screen or angled filter to allow water to pass through while diverting larger debris away.
- Prevent clogs in rainwater storage systems, keeping sediment and organic matter out.

Benefits of Downspout Filters (Leaf Eaters)

- Reduces maintenance and prevents buildup in storage tanks, prolonging system lifespan.
- Improves water quality by keeping out leaves, insects, and debris that can cause contamination.
- Prevents clogs and protects gutters, downspouts, and pipes from blockages.
- Enhances filtration efficiency by working as the first step in a multi-stage rainwater filtration system.

Feedlines (Inflow Lines)

A feedline, also called an inflow line, is the pipe or conduit that transports collected rainwater from the downspout to a storage system, such as a rain barrel, cistern, or underground tank. It ensures that rainwater is directed efficiently and safely into the storage unit while minimizing debris and contamination.

Key Functions of a Feedline/Inflow Line

- Channels water from gutters and downspouts into a storage system.
- Helps control water velocity to prevent excessive splashing or erosion at the entry point.
- Can be gravity-fed or pressurized (when using a pump system).
- Works with pre-filtration systems like first-flush diverters or downspout filters to remove debris before water enters storage.

Types of Feedlines/Inflow Lines

- **Direct Piping:** Uses PVC, HDPE, or flexible tubing to connect downspouts to storage tanks.
- **Open Troughs or Channels:** Sometimes used in large-scale or agricultural setups to guide water into reservoirs.
- **Subsurface Piping:** In underground rainwater storage systems, inflow lines may be buried to discreetly channel water into cisterns.

Best Practices for Feedlines

- Use non-toxic, durable materials (e.g., Schedule 40 PVC, food-grade hoses) to prevent contamination.
- Ensure a slight downward slope for gravity-fed systems to maintain smooth water flow.
- Integrate pre-filtration (e.g., mesh screens, first-flush diverters) to remove sediment and organic matter before storage.
- Position the inlet properly to prevent turbulence or overflow inside the tank.

Distribution Lines

Distribution lines are the pipes, hoses, or channels that transport stored rainwater from the collection tank or cistern to its final use, such as irrigation, household non-potable use, or groundwater recharge. These lines are essential for efficiently and effectively delivering harvested rainwater where it's needed. One of the key decisions when designing a rainwater harvesting system is whether to use a pump-driven or gravity-fed system to distribute the stored water. Each method has advantages and limitations, depending on your storage size, elevation, and water pressure needs.

Types of Distribution Lines

Gravity-Fed Distribution Lines

- Used in simple, low-pressure systems where water flows naturally due to elevation.
- Typically made of PVC, polyethylene tubing, or garden hoses.
- Best suited for drip irrigation, soaker hoses, and slow-release watering systems.
- Requires correct slope and elevation to maintain consistent water flow.

Pump-Driven Distribution Lines

- Used in pressurized systems where water must be delivered with force.
- Can handle longer distances and uphill transport of water.
- Often made of reinforced PVC, PEX tubing, or HDPE pipes to withstand higher pressure.
- Ideal for sprinkler systems, hose attachments, and indoor non-potable uses (toilets, laundry, etc.).

Subsurface and Infiltration Lines

- Designed to recharge groundwater or deliver water directly into the soil.
- Includes perforated pipes, French drains, and underground irrigation lines.
- Helps distribute excess rainwater into rain gardens, bioswales, or infiltration basins to prevent runoff and erosion.

Key Considerations for Distribution Lines

- **Material Selection:** Use UV-resistant, non-toxic piping to prevent degradation and contamination.
- **Proper Sizing:** Match pipe diameter to flow rate and pressure needs for efficiency.
- **Filtration and Maintenance:** Keep lines free of debris using filters and regular flushing to prevent clogs.
- **Flow Control:** Use valves, timers, and pressure regulators to control water distribution effectively.

Hydrants

A hydrant, or frost-free hydrant, is an outdoor water access point that connects to a rainwater storage system, allowing for easy, controlled water distribution. It functions like a traditional water spigot but is designed to operate in all weather conditions, including freezing temperatures.

Water is carried from a pump via ¾" (1.9 cm) PEX lines to yard hydrants, which should be frost-free and secured to posts.

GRAVITY-FED SYSTEMS: SIMPLE, PASSIVE, AND LOW-MAINTENANCE

A gravity-fed rainwater system relies on elevation and natural downward flow to move water from a storage tank to its point of use. This method is cost-effective, energy-efficient, and requires little maintenance, but it has lower water pressure compared to pumped systems.

Key Considerations for Gravity-Fed Systems:

- Ideal for low-volume uses like drip irrigation, hand watering, or slow-release watering systems.
- Works best when the tank is elevated above the point of use to generate pressure.
- Elevation matters—the higher the tank, the stronger the water pressure. Every 2.3 feet (70 cm) of elevation provides 1 PSI (pound per square inch) of pressure.
- Water moves more slowly than in a pumped system, making it less effective for high-demand applications like hose attachments or household plumbing.
- Tanks must be securely placed on a reinforced platform to support the heavy weight of water (8 pounds per gallon [1 L per kg]).

Pump-Driven Systems: Higher Pressure for Greater Versatility

A pump-driven system uses electric, solar, or manual pumps to increase water pressure, allowing for more consistent and higher-volume water distribution. This is recommended for tanks over 300 gallons (1,136 L) or when pressure needs exceed what gravity alone can provide.

Key Considerations for Pump Systems

- Provides strong, consistent pressure for garden hoses, irrigation systems, and household use.
- Necessary for pressurized irrigation systems, such as sprinklers or high-volume drip irrigation.
- Works at any elevation, meaning tanks don't need to be placed high up for functionality.
- Requires electricity or solar power, adding operational costs and maintenance needs.
- Some pumps include pressure regulators and automatic shutoff features for efficiency and safety.

Which One Should You Choose?

Use a gravity-fed system if you want a low-tech, low-cost, energy-free option for slow irrigation or hand watering.

Use a pump system if you need stronger water pressure, higher flow rates, or the ability to connect to hoses, sprinklers, or household fixtures.

Hybrid systems can use gravity for low-flow applications while integrating a pump for higher-pressure needs when necessary.

Above This cistern collects rainwater for irrigation, and overflows via gravity to nearby rain gardens.

How Hydrants Work in a Rainwater Harvesting System

- Installed as an outlet for cisterns, tanks, or underground rainwater storage systems.
- Provides manual access to harvested rainwater for irrigation, livestock watering, or general outdoor use.
- Can be gravity-fed (if the tank is elevated) or pump-assisted for increased pressure.

Frost-Free Hydrants: Why They Matter

- A frost-free hydrant is designed to prevent freezing in colder climates by incorporating a drain-back system.
- The water supply shuts off below the frost line (typically 3 to 4 feet [0.9 to 1.2 m] underground).
- Any remaining water in the standpipe drains out, preventing ice buildup and pipe bursts.
- Ensures year-round functionality for rainwater access, even in winter.

Benefits of Using a Hydrant in Rainwater Harvesting

- Provides easy access to stored rainwater for hoses, buckets, or irrigation systems.
- Eliminates freezing risk in cold climates (frost-free models).
- Durable and long-lasting, with heavy-duty materials like galvanized steel or brass.
- Can be integrated into gravity-fed or pump-driven systems for various water pressure needs.

Best Uses

- Garden and landscape irrigation (connects to hoses or drip systems).
- Livestock watering in rural or off-grid settings.
- Emergency water supply access from large storage tanks or cisterns.
- Outdoor cleaning and maintenance (washing equipment, vehicles, patios, etc.).

HOW WINTER DRAINS WORK IN INVERTED SIPHON SYSTEMS

An inverted siphon system is a U-shaped pipe configuration designed to move water from one elevation to another, often under obstacles like roads, pathways, or structures. Because these systems create natural low points where water can collect, a drain valve or auto-drain system must be installed at the lowest point to allow for complete evacuation of water before freezing temperatures set in.

Winter Drains

Winter drains are essential for inverted siphon systems and other water transport lines in cold climates, where freezing temperatures can cause water to expand and burst pipes or damage components. These drains allow water to fully evacuate from the system, preventing ice buildup and ensuring smooth operation when temperatures drop.

Why Are Winter Drains Necessary?

- **Prevent freezing damage.** Water left in pipes during freezing conditions can expand and crack pipes, valves, and fittings. A properly placed drain eliminates standing water in low points where it might freeze.
- **Ensure system longevity.** Regularly draining pipes before winter protects the infrastructure and reduces costly repairs.
- **Keep the system functional in spring.** Without ice-related damage, the rainwater harvesting system can resume operation immediately when temperatures rise.

INVERTED SIPHON SYSTEM IN RAINWATER HARVESTING

An inverted siphon system in rainwater harvesting is a specialized piping configuration that moves water under obstacles or across elevation changes without requiring a pump. Instead of allowing gravity to carry water downhill in an open channel, an inverted siphon directs water downward to a low point and then back up to a higher elevation on the other side, using water pressure and gravity.

When installing rainwater storage tanks, placement is key. Because water seeks its own level, tanks don't always need to be directly next to a catchment source like a roof. Instead, they can be positioned away from a structure or even slightly uphill as long as the inflow line maintains the proper elevation change (e.g., a 2-foot rise over 100 feet [61 cm rise over 30.5 m]).

How It Works

- Water enters the system at a higher elevation (e.g., from a downspout or cistern).
- It flows downward to a low point (e.g., beneath a driveway, walkway, or other obstruction).
- Water pressure and siphoning action push it back up to a discharge point at or near the original elevation.
- The system ensures that water reaches areas where a direct downhill flow isn't possible.

Uses of Inverted Siphons in Rainwater Harvesting

- Routing water under roads, pathways, or structures without disrupting the landscape.
- Transferring collected rainwater between different levels of a property (e.g., from a roof catchment system to a distant or slightly elevated storage tank).
- Managing overflow by directing excess water to designated infiltration areas such as swales or rain gardens.
- Balancing multiple storage tanks that are positioned at different elevations.

Key Considerations for Using an Inverted Siphon in Rainwater Harvesting

- **Proper pipe sizing:** Ensures smooth water flow without excessive resistance.
- **Avoiding airlocks:** Venting or vacuum breakers may be needed to prevent trapped air from stopping water flow.
- **Winterization:** A drain valve at the lowest point prevents freezing in cold climates.
- **Regular maintenance:** Debris can settle in the low point, so periodic flushing is essential.

EFFECTIVELY MANAGING RAINWATER HARVESTING OVERFLOW

To prevent flooding and erosion, rainwater harvesting systems should be designed with overflow management in mind. When a tank reaches capacity, excess water must be directed safely into the landscape. I recommend designing rainwater harvesting systems assuming that the tank is always full.

- Use contour swales or rain gardens to distribute overflow evenly and allow water to infiltrate into the soil.
- Ensure proper grading (e.g., 2-foot drop over 100 feet [61 cm drop over 30.5 m]) to guide overflow to a receiving area without causing erosion or water pooling.
- Integrate earthworks such as check dams, retention basins, or infiltration trenches to handle large overflow volumes effectively.
- When siting tanks, consider water levels: Water seeks its own level, so tanks can be placed away from the house or even uphill as long as the inflow line is properly elevated (e.g., 2-foot rise over 100 feet [61 cm over 30.5 m]).

Integrate overflow into your landscape using contour swales or rain gardens. Design earthworks to handle overflow as if your tank is full and ensure proper grading (2-foot drop over 100 feet [61 cm over 30.5 m]) for water flow to the areas receiving the overflow.

Calculating Water Needs and Storage

HOW MUCH WATER DO YOU NEED TO CAPTURE FOR IRRIGATION?

Understanding how much water your landscape requires is key to designing an efficient and sustainable irrigation system. A general rule of thumb is that most landscapes thrive with 1 inch (2.5 cm) of water per week, but how you deliver that water can make a big difference in water efficiency, plant health, and soil moisture retention.

Calculating Your Water Needs

To simulate 1 inch (2.5 cm) of rainfall per week, use the following formula:

Square footage of the landscape × 0.622 = gallons needed per 1 inch (2.5 cm) of water where 0.622 is the conversion factor that determines how many gallons cover 1 square foot with 1 inch of water (sq m × 10 = L per 2.5 cm of water).

For example:
A 1,000 square-foot garden needs 622 gallons per week (A 93 sq m garden needs 2,325 L per week).
A 5,000 square-foot lawn needs 3,110 gallons per week (A 465 sq m garden needs 11,625 L per week).

How Much Water Can You Capture?

To calculate how much water-capture potential your roof has, use the following formula:

Square footage of the roof × 0.622 × annual rainfall (inches) = potential gallons captured per year (sq m × 10 × annual rainfall [cm] = potential L).

For example:
A 1,000 sq ft roof with 50 inches of annual rainfall can capture 31,100 gallons per year (A 93 sq m roof with 127 cm of annual rainfall can capture 118,110 L per year).

A 5,000 sq ft roof with 50 inches of annual rainfall can capture 155,500 gallons per year (A 465 sq m roof with 127 cm of annual rainfall can capture 590,550 L per year).

Once you know your total potential rainwater you are able to harvest, next consider how many weeks in a year you might go without 1 inch (2.5 cm) of rain on your site. In Atlanta, our longest recent drought during the growing season was about twelve weeks. On average, however, we usually go three to four weeks in a summer without rain.

If you wanted to create a rainwater capture system to irrigate a 200 square-foot (19 sq m) garden, here is how you would figure out your storage needs for four weeks of drought:

200 sq ft × 0.622 = 124 gallons of water needed per week to simulate 1 inch of rain × 4 weeks = 496 gallons of storage needed to irrigate for 4 weeks during a drought (19 sq m × 10 × 2.5 cm × 4 weeks = 1,878 L of storage needed to irrigate for 4 weeks during a drought).

You would then evaluate the storage needs against the potential rainwater you are able to harvest from your catchment area. Let's say you are harvesting from 1,000 sq ft (93 sq m) of roof. Your total potential water in a year is 31,100 gallons (118,110 L) of water, so your catchment area is more than sufficient to accommodate your irrigation needs.

CHOOSE EFFICIENT IRRIGATION METHODS

Once you determine how much water your landscape needs, select an irrigation system that minimizes waste and maximizes absorption.

Drip Irrigation

- Provides water directly to plant roots, minimizing evaporation and maximizing efficiency.
- Best for gardens, food crops, trees, and shrubs.
- Can be connected to rainwater harvesting systems for sustainable watering.

Soaker Hoses

- Slowly release water along the hose, allowing deep soil penetration.
- Ideal for mulched beds, perennial gardens, and vegetable rows.

Mulching

- Helps retain moisture, reducing irrigation needs.
- Use straw, wood chips, or compost around plants to prevent evaporation.

Rainwater Harvesting Integration

- Use cisterns, rain barrels, or swales to collect and store water.
- Gravity-fed or pump-assisted irrigation can distribute rainwater efficiently.

Smart Scheduling

- Water early in the morning or late in the evening to prevent evaporation.
- Adjust irrigation based on seasonal rainfall, soil moisture, and plant needs.

Deep, Infrequent Watering

- Instead of shallow daily watering, apply water less frequently but deeply to encourage strong root growth and drought resilience.

With water management strategies in place, the next step in building a regenerative landscape is focusing on the foundation of all plant life: soil. Soil is more than just dirt—it's a living system that supports biodiversity, retains moisture, and cycles nutrients. Unfortunately, conventional landscaping practices degrade soil health, leading to erosion and nutrient loss. Section 2 will show you how to build and maintain healthy, fertile soil using natural techniques that mimic the cycles of a thriving ecosystem.

Above This tidy veggie garden pairs lush, productive beds with pervious gravel paths that let rain soak in.

THREE COMMON MISTAKES WITH RAINWATER HARVESTING (AND HOW TO AVOID THEM!)

When it comes to rainwater harvesting, a few common mistakes can leave the end product less than functional. I'm going to walk you through those errors and show you how to avoid them, so you have a beautiful and functional rainwater system.

MISTAKE #1: The cistern or barrel is not sized appropriately

An immense amount of water can come off of a standard residential roof. The number one mistake I see in the field is that the tanks are undersized for the amount of water they collect. As you can imagine, a small rain barrel collecting half of your roof water is going to fill up VERY quickly, and then, more often than not, the water pours out of the top (usually right at the foundation). This leads me to the second error I see often.

MISTAKE #2: The cistern or barrel doesn't have an overflow (or if it does, it's too small)

Especially when you're working with a small rain barrel, it's essential to make sure you have a properly sized overflow. Most rain barrels come equipped with a 1-inch (2.5 cm) overflow. But think about this: If the barrel is full (and you should always plan as if it is because often, it will be), and then a full 3-inch (8 cm) downspout is pouring into the tank that quickly gets sized down to a 1-inch (2.5 cm) pipe (the manufacturer's overflow). This is a textbook definition of a bottleneck, and oftentimes, the water ends up spilling out of the top of the tank at the foundation (again!). A good rule of thumb is: The overflow should be the same size as the input pipe (in most cases, the downspout that feeds it).

MISTAKE #3: There's nowhere for the water to go

So, what do you do once you have an overflow? Send it somewhere useful!

The third error I often see with rainwater harvesting is after having gone through all the trouble of hooking up a rain cistern, the overflow (if there is one) either dumps out right next to the tank (again near the foundation!) or it is sent down the driveway, lawn, or other impervious surface and into the stormwater system. And yes, I know I called lawn an impervious surface—did you know that between 80 to 90 percent of the water that hits the lawn runs off (depending on the type of grass)? That's another topic altogether . . . Instead, there is a great opportunity literally pouring out of that overflow pipe to be put to service restoring the water cycle in your yard! You can plumb it to go to a rain garden, swale, or even a heavily mulched garden bed, where it can soak into the soil.

Above A permeable path helps water infiltrate, rather than creating more runoff.

Fungi growing from leaf litter on the forest floor.

Section 2

Soil

Pillar #2: Integrate "Waste" as a Resource to Build Soil Fertility and Create Resilience

Beneath our feet lies the most important yet often overlooked component of a thriving ecosystem: soil. More than just dirt, soil is a living, breathing system—a vast underground web of microbes, fungi, and organic matter that supports all plant life. Healthy soil holds water like a sponge, cycles nutrients efficiently, and serves as a foundation for biodiversity. Yet, modern landscaping and agriculture have degraded this precious resource through overtilling, chemical inputs, and soil compaction, leaving behind dry, lifeless dirt that requires constant intervention to sustain plant growth.

In this section, we'll dive into the science of soil health and how to restore its natural fertility. You'll learn about the harmful impacts of conventional soil practices and how to build a circular economy in your garden, where organic waste becomes nourishment rather than landfill. We'll explore regenerative techniques like *hügelkultur*, lasagna gardening, and compost tea—methods that enrich the soil, store carbon, and create self-sustaining fertility with minimal effort.

By shifting the way we think about soil, we can move from depletion to regeneration. Instead of stripping the land of nutrients and relying on synthetic fertilizers, we can nurture a rich, living soil that feeds itself, and in turn, feeds everything growing in it. Whether you're starting with compacted clay, sandy soil, or nutrient-depleted beds, the strategies in this section will help you rebuild the foundation of your landscape—one that is resilient, productive, and teeming with life.

"To forget how to dig the earth and to tend the soil is to forget ourselves."

MAHATMA GANDHI

Soil nutrients often end up in our waterways as runoff, causing sedimentation of our rivers and creeks while depleting soil fertility.

CHAPTER 4

THE PROBLEM WITH HOW WE TREAT OUR SOILS

Where do our soil nutrients go? The short answer is: into our waterways and into landfills. Most landscapes are managed according to the principles of a linear economy, where resources are extracted, made into something else, and then disposed of off-site. This requires a constant stream of extraction and waste in order to function. However, resilient landscapes create closed-loop, circular economies in the garden because there is no "waste" in nature.

In this chapter, we'll explore the second pillar of a regenerative landscape: soil. This essential component of sustainable landscaping focuses on understanding how common practices can undermine soil health and identifying site conditions that may hinder nutrient management and the success of your future landscape.

What Constitutes Healthy Soil?

Healthy soil is a living, dynamic system that supports plant life, retains water, and cycles nutrients efficiently. It is teeming with microbial life, organic matter, and beneficial fungi, which work together to create a self-sustaining environment that promotes plant growth and ecosystem resilience.

KEY CHARACTERISTICS OF HEALTHY SOIL

High Organic Matter Content

- Contains decomposed plant material, compost, and humus, which provide nutrients and improve soil structure.
- Enhances water retention and nutrient availability for plants.
- Supports soil microbes and earthworms, which break down organic matter into plant-accessible nutrients.

Balanced Soil Structure (Aggregates and Pores)

- Loamy texture (a mix of sand, silt, and clay) that holds water while allowing air to circulate.
- Well-formed soil aggregates (clumps of soil particles) provide stability, prevent erosion, and create space for root growth.
- Porous and aerated, allowing roots to penetrate deeply and water to infiltrate effectively.

Thriving Microbial and Fungal Communities

Healthy soil is alive with bacteria, fungi (such as mycorrhizal networks), and other microorganisms that:

- Fix nitrogen, making it available for plant uptake.
- Break down organic material into nutrients.
- Improve plant resistance to diseases and pests.

Water Infiltration and Retention

- Absorbs and stores rainwater, reducing runoff and preventing erosion.
- Maintains moisture levels for longer periods, supporting plants during dry spells.
- Prevents compaction, which allows roots and organisms to move freely.

Nutrient Cycling and pH Balance

- Contains adequate levels of nitrogen, phosphorus, potassium, and trace minerals.
- A balanced pH (typically between 6.0 and 7.5) allows plants to absorb nutrients effectively.
- Naturally replenishes nutrients through decomposition, root exudates, and microbial interactions rather than relying on synthetic fertilizers.

HOW HEALTHY SOIL CONTRASTS WITH SOILS IN DEVELOPED URBAN AREAS

In contrast, urban soils are often degraded, compacted, and stripped of their natural biodiversity due to construction, pollution, and poor land management. Here's how urban soils differ from healthy soils.

Compacted and Disturbed Structure

- Heavy machinery, paving, and construction cause soil compaction, reducing porosity and making it difficult for roots, air, and water to penetrate.
- Loss of soil layers (topsoil removal) leads to poor plant growth and reduced water retention.
- More prone to erosion and runoff, as rainwater cannot infiltrate easily.

Low Organic Matter and Microbial Life

- Urban soils often lack organic material due to removal of plant life, leaves, and natural decomposition cycles.
- Limited soil microbes and fungi, resulting in poor nutrient cycling and weak plant resilience.
- Overuse of synthetic fertilizers and pesticides can further degrade microbial diversity.

Poor Water Retention and Increased Runoff

- Urban soils are often hydrophobic (water-repellent) due to compaction and lack of organic matter.
- Water runs off quickly, leading to flooding, erosion, and poor groundwater recharge.
- High levels of impervious surfaces (concrete, rooftops, asphalt) prevent infiltration entirely, increasing stormwater issues.

Contaminants and Chemical Imbalances

- May contain heavy metals, construction debris, synthetic chemicals, and pollutants from urban runoff.
- Unbalanced pH due to industrial waste or excessive chemical applications, making it harder for plants to absorb nutrients.
- Often lack key minerals and nutrients, requiring constant amendments for plant survival.

Studies in the *Journal of Environmental Quality* suggest that synthetic nitrogen fertilizers contribute to soil nitrogen depletion, which can result in erosion, runoff, and contamination of water sources. This decline in soil health also reduces its ability to store carbon dioxide, ultimately playing a role in climate change.

When fertilizer runoff reaches streams, lakes, and oceans, it creates a cascade of environmental harm. Nitrogen and phosphorus from fertilizers trigger algal blooms, which initially thrive but eventually decompose, depleting oxygen in the water. This process, known as eutrophication, creates "dead zones"—areas so oxygen starved that aquatic life cannot survive. In 2008, over four hundred dead zones were identified globally, and in 2016, one in the Gulf of Mexico had grown to the size of Connecticut due to nitrogen runoff.

Using synthetic fertilizers, pesticides, and herbicides has severe consequences for both local landscapes and the broader environment. These chemicals don't just eliminate pests and weeds—they disrupt the delicate balance of soil ecosystems, killing beneficial organisms that are essential for carbon sequestration, soil health, and plant resilience. Over time, repeated chemical use depletes soil fertility, making it more hospitable to invasive species while creating a cycle of dependence on artificial inputs. As soil quality declines, maintaining plant health becomes increasingly expensive and unsustainable—a burden no homeowner wants to take on.

Beyond environmental damage, the health risks associated with these chemicals are well documented. Numerous studies have linked chemical exposure to serious diseases, and growing legal actions, bans, and record-breaking settlements highlight the dangers of conventional landscaping and agricultural chemicals. Is their use worth the risk?

The impact doesn't stop at individual gardens—it extends to entire ecosystems. Chemical runoff contaminates water sources, disrupts pollinator populations, and reduces biodiversity, setting off a chain reaction of ecological decline. Through biomagnification, these toxic compounds accumulate in food chains, worsening their effects on wildlife and humans alike. Additionally, by reducing carbon sequestration and increasing greenhouse gas emissions, these chemicals actively contribute to climate change, making our environment less resilient to extreme weather and natural disasters.

The interconnectedness of our ecosystems means that every choice matters. Shifting away from synthetic inputs not only protects soil, water, and biodiversity but also supports a more sustainable, self-sufficient landscape that thrives without the need for harmful chemicals.

When we consider rampant herbicide and pesticide use in landscapes today, it is really a symptom of a larger, systemic issue. Most landscapes today are managed according to a linear economic model: Resources are extracted, used, and discarded. This system depends on a continuous cycle of extraction and waste, in stark contrast to nature's approach, where no waste exists.

A CASE STUDY IN LINEAR PRACTICES: LEAF MANAGEMENT

Consider the conventional method of leaf management. Leaves are blown away using fossil-fuel-powered equipment, gathered into plantation-grown-pine-pulp or plastic bags, and hauled off to landfills on fuel-guzzling trucks. Come spring, gardeners then purchase composted leaves, often shipped from distant locations. This inefficient cycle depletes local ecosystems and contributes to climate change. According to the United States Environmental Protection Agency (EPA), Americans sent approximately 14.4 million tons (13.1 million metric tons) of yard trimmings to landfills in 2011. When organic material decomposes in landfills without access to oxygen, it generates methane, a powerful greenhouse gas. The EPA notes that "removing such biomass from its original site reduces the potential for carbon sequestration in the soil."

This approach exemplifies a broken loop, a hallmark of the linear economy. By hauling away organic matter, we create a need to import it back, perpetuating waste and inefficiency.

Above Bags of leaves in the autumn, waiting to be hauled away.

ORGANIC MATTER MATTERS

Soil is more than just dirt—it's a living, breathing ecosystem. At its foundation, about half of it is made up of air and water, while around 45 percent consists of minerals like sand, silt, and clay. The remaining 5 percent is organic matter, made up of decomposing plants and animals. Even though this organic matter is a small part of the mix, it plays a huge role in soil health. It feeds microorganisms that help store water, delivers nutrients to plants, and even keeps pests in check.

Soil isn't the same everywhere—it changes with climate and conditions. But no matter where you are, there's always an essential relationship between plants, fungi, bacteria, and other tiny organisms that keep the soil thriving and full of life. Building soil through the addition of organic matter has numerous impacts in your garden and beyond. Pinpointing the exact depth of topsoil in the eastern United States before European settlement is difficult due to a lack of historical records. However, it is widely recognized that the region once had exceptionally fertile

soil, particularly in ecosystems like the tallgrass prairies, where deep layers of topsoil were prevalent. Over time, widespread agriculture and deforestation have contributed to extensive soil erosion and depletion.

The rate at which topsoil naturally forms depends on environmental factors, with estimates varying significantly. The Minnesota Association of Soil and Water Conservation Districts reports that it can take anywhere from thirty to one thousand years for just one inch of topsoil to develop. Given this slow regeneration process, adopting sustainable land management practices is essential for preserving and rebuilding soil health.

To begin building soil fertility, we first need to identify the sources of organic matter available both on your site and within your community. Organic matter can be divided into two main categories: nitrogen sources and carbon sources. Common nitrogen sources include green yard waste, grass clippings, kitchen scraps, and even urine for those adventurous enough to use it. On the other hand, carbon sources often come from leaves, branches, wood chips, sawdust, cardboard, and paper waste.

The Soil in Your Landscape

Identifying and understanding the soil on your site is fundamental when creating a permaculture design because soil is the foundation of all life and productivity in a landscape. Healthy soil supports plant growth, retains water, and fosters biodiversity, creating a resilient ecosystem. By assessing soil composition, structure, and nutrient cycles, you can strategically enhance fertility, improve water retention, and prevent erosion, ensuring long-term sustainability and abundance in your landscape.

WHAT SHOULD YOU LOOK FOR WHEN EVALUATING YOUR SOIL?

In sub/urban landscapes, several common site conditions indicate opportunities for better nutrient management:

- **Erosion and loss of topsoil:** Signs of soil erosion suggest a need to stabilize and retain nutrients.
- **Lack of organic matter:** Insufficient organic content hinders soil fertility and structure.
- **Compaction:** Dense, compacted soil reduces water infiltration and root growth, limiting productivity.

You can evaluate the following conditions on your site and make notes about what you are observing as part of your site analysis.

Leaf litter makes an excellent mulch as it breaks down into rich organic material. Adding leaves to your perennial plantings, such as with the yarrow (*Achillea millefolium*) featured here, is a great way to suppress weeds.

Soil Profile

All soil is made up of clay, sand, and silt, which refer to the particle size. You can do a simple jar test if you are curious about the composition of your soil. To do that, add a small scoop of your soil (from the top layer to the subsoil) into a jar and fill it with water. Shake the jar and let it sit overnight. The particles will settle and stratify, and you will be able to see the ratio of sand (on the bottom), silt (in the middle), clay (on the top), and organic matter (floating). Over time, as you build organic matter, the profile may shift somewhat. You'll read that 30:30:30 is the "ideal" ratio, but that's not necessarily true. In very few places does that ratio exist in nature. What *is* ideal is knowing your soil profile and choosing plants and strategies that utilize your soil's innate characteristics as an asset, so you are working with nature rather than trying to change it.

Percolation Rate

Soils with more sand will allow water to infiltrate more quickly. The higher the clay content, the slower the water will percolate. Building organic matter in the soil helps to equalize the infiltration rate. If you have very heavy clay soils, building organic matter helps more water infiltrate. If you have very sandy soils, organic matter helps hold more moisture. Either way, you can probably see a trend: Build organic matter!

Exposed Soil

Remember that it is ideal to keep all soil covered, so observe your site and make note of any areas that show bare, exposed soil. Make note of the site conditions that may be causing it. Is water runoff carrying away your topsoil? Or is it a high traffic area made bare through use? Or perhaps it's an area under a thick-canopied tree that receives little water and light?

Compaction

How compacted is your soil? Is it fairly easy to dig into? Are there areas that receive high foot traffic or vehicle traffic? Water running off, rather than infiltrating, can also lead to compaction because dehydrated soils have less air space for roots and organic material.

Topsoil

Do you have much topsoil? You can determine the amount by digging beneath the mulch layer or leaf litter to see if there is an assimilation layer in your soil, where the larger mulch layer is beginning to decompose.

Topographical Position

What's your topographical position? Ridge, valley, mid-slope? Valleys tend to have more alkaline soils because they collect sediment, topsoil, and nutrients that are carried in water from uphill. Soils lower in the watershed also tend to be more mesic. Alternatively, there tend to be more acidic soils toward ridges and along slopes. Many nutrients are carried out of the soil with erosion, leading to less organic matter toward ridge tops. Soils higher in the watershed also tend to be dryer.

Pests and Diseases

Pests and diseases, such as molds, mildews, blights, and pests, often indicate poor soil health. Make sure the plants you introduce are well suited for the moisture and light conditions in your garden and build your soil microbiome by encouraging the growth of beneficial bacteria and microorganisms, which increase nutrient availability.

Regenerative Soil Strategies

Regenerative landscapes embrace a circular economy, where waste becomes a resource. By adopting soil-building techniques tailored to your site's conditions, you can retain nutrients, improve fertility, and enhance resilience. Through thoughtful management and a shift away from the linear model of resource management, we can create landscapes that nurture the soil, sustain biodiversity, and contribute to climate solutions.

Conventional gardening and agriculture treat soil as an expendable resource, often depleting its fertility through chemical inputs and overtilling. But healthy soil isn't something we have to constantly replace—it's something we can regenerate. In chapter 5, we'll explore how to create a circular economy in your garden, where organic matter is continuously recycled to build soil fertility, reduce waste, and enhance ecosystem resilience.

Part habitat, part sculpture—this birdhouse made from reclaimed wood and a dead tree limb is a beautiful example of ecological art. It celebrates nature, reuse, and the wildlife that call our gardens home.

This garden illustrates a circular economy by using repurposed materials, integrating waste as a resource, and embracing the natural forces of succession at play.

CHAPTER 5

HOW TO BUILD A CIRCULAR ECONOMY IN YOUR GARDEN

A thriving garden is more than just a collection of plants—it is a dynamic system where resources cycle continuously to sustain soil health, biodiversity, and long-term productivity. A circular economy approach in the garden eliminates waste, repurposes organic materials, and regenerates natural systems, reducing reliance on external inputs while enriching the land. In this chapter, you'll learn how to apply the principles of a circular economy to your landscape by designing out waste, keeping materials in use, and supporting natural regeneration. Through composting, mulching, and fostering ecological interactions, you'll discover practical strategies to build healthier soil, reduce waste, and create a more self-sustaining garden.

What Is a Circular Economy?

Creating a circular economy in your garden is an excellent way to accelerate soil health. A circular economy is built on three principles: Design out waste, regenerate natural systems, and create closed-loop systems. In the case of soil, this can be achieved by composting yard waste and kitchen scraps, using cardboard for sheet mulching, and regenerating natural systems by attracting wildlife and supporting nutrient cycling. This approach helps reduce the need for external inputs and keeps materials in use, ultimately benefiting both the soil and the environment. A circular economy in this context means continuously cycling resources—minimizing waste while maximizing natural processes of regeneration.

DESIGN OUT WASTE AND POLLUTION

Pollution occurs when something accumulates in such a high concentration that it disrupts ecosystem function. In soil management, we can reduce waste by:

- Composting yard waste and kitchen scraps, returning nutrients to the soil instead of sending them to landfills.
- Keeping leaves on-site, using them as mulch in planting beds rather than removing them as "waste."

KEEP PRODUCTS AND MATERIALS IN USE

By repurposing organic materials, we can enrich soil and reduce dependency on external inputs.

- Sheet mulching with cardboard from shipping boxes creates a weed barrier and builds organic matter.
- Shredded paper can serve as a cover material in compost piles, balancing nitrogen-rich food scraps with carbon sources.

REGENERATE NATURAL SYSTEMS

Healthy soil ecosystems mimic natural nutrient cycling, much like a forest floor, where organic matter is continuously broken down and replenished. We can support this process by:

- Allowing leaves to decompose in place, returning nutrients to the soil.
- Attracting natural fertilizers by welcoming wildlife and birds whose waste enriches the soil.
- Using companion planting to enhance soil health.
- Maintaining intact forested areas, which serve as nutrient reservoirs and soil builders.
- Creating closed-loop systems, such as integrating chickens into the garden cycle: Their manure enriches compost, and in warmer winter climates, they graze garden beds, naturally fertilizing the soil.
- Covering crop areas and maintaining soil cover to protect against erosion, retain moisture, and build long-term fertility.

By following these principles, we shift from extracting resources to regenerating them, building soil resilience, fertility, and self-sustaining abundance—all while reducing waste and closing the loop on nutrient cycles.

How to Create Healthy Soils

Soil health is the foundation of a thriving, resilient landscape. In regenerative landscaping, the goal is to create self-sustaining ecosystems that increase biodiversity, improve water retention, and build long-term soil fertility. Healthy soil supports all aspects of the landscape by enhancing plant growth, improving water management, and sequestering carbon, making it a critical component of ecological land management.

Healthy soil is not just dirt—it is a living, dynamic system that supports resilient landscapes, retains water, reduces carbon emissions, and sustains biodiversity. Soil plays a critical role in building a regenerative landscape.

As mentioned, soil composition varies based on climate, yet all soil ecosystems rely on a complex network of interactions between plants, fungi, bacteria, and other organisms. Enriching soil with organic matter can have widespread benefits, improving both local garden health and broader environmental conditions.

Additionally, soil is a key player in the global carbon cycle. Plants absorb carbon from the atmosphere and transfer it to the soil through their roots, nourishing microbes that, in turn, supply plants with essential nutrients. However, when soil is disturbed, stored carbon is released as CO_2, contributing to greenhouse gas emissions. Disruptive practices like tilling can degrade organic matter and weaken soil structure, making soil health a critical factor in mitigating climate change.

PROMOTING SOIL HEALTH THROUGH REGENERATIVE PRACTICES

Keep the Soil Covered with Plants

You can do this a few ways. Establish a living mulch of ground cover throughout your planting areas. Use triple-ground fine wood chips or composted leaf mulch to encourage the quick spread of herbaceous plants, as opposed to chunky wood chips to suppress weeds.

If you're gardening, cover crop areas when you're not using them for growing. Cover cropping is the

Top Covering the ground with plants reduces erosion and increases soil carbon and its water-retention capacity.

Bottom Planting a diversity of species aids the soil profile through different root structures and functions and diversifies the pollinators and wildlife your landscape will attract.

practice of seeding bare ground to prevent erosion, improve water infiltration, and build organic matter. It's a valuable strategy for preparing the soil when you're not yet ready to plant.

The choice of cover crop depends on the season and site conditions. For winter, winter rye offers deep roots for soil decompaction, while Austrian winter pea serves as a nitrogen fixer and animal forage. Spring and fall are ideal for hairy vetch, a quick-growing nitrogen fixer, and daikon radish, which helps break up compacted soil. Summer options include buckwheat for pollinator support and alfalfa for enhancing soil biota. At the end of the growing season, you can employ chop and drop—cutting the cover crop at the base and leaving it to decompose, enriching the soil for future plantings.

Ensuring that the soil is always covered with plants is one of the most effective ways to build soil health. It also helps prevent erosion, increase carbon storage, and improve the soil's ability to retain water.

Diversify What the Soil Grows

Grow many layers of plants, including ground covers, forbs, brambles, shrubs, and trees and plant densely. If you're working with a tight budget, choose species that are easy to propagate by seed or divisions, or choose species that spread quickly but aren't invasive. Certain plants are "heavy feeders" meaning they use heavy loads of nutrients in the soil to build their aerial parts, which if grown as a monoculture can deplete the soil. Diversifying your plantings within an area helps replenish soil nutrients and bolsters the health

GROWING MUSHROOMS AT HOME

Growing mushrooms at home is an easy, rewarding, and inexpensive way to produce nutritious food while learning about fungal cultivation. And in the spirit of a circular economy, using logs from trees you may have to remove for any number of reasons is a great way to keep the resources in circulation! Whether you choose to grow mushrooms on logs or in a sawdust-based system, this guide will walk you through the process step by step.

Shiitake spores inoculate logs using the dowel method.

Materials Needed

- → Fresh logs (3 to 5 inches [8 to 13 cm] in diameter, about 4 feet [122 cm] long)
- → Inoculated dowels (purchased from a mushroom supplier)
- → Drill (bit size matches dowels)
- → Hammer or mallet
- → Beeswax or cheese wax
- → Double boiler (for melting wax)
- → Paintbrush (for sealing holes)
- → Shady, moist outdoor space for incubation

METHOD 1: GROWING MUSHROOMS ON LOGS

Step Instructions

1. **Source Your Logs**
   - Contact a local tree service to obtain freshly cut hardwood or softwood logs. Choose the right wood:
     - **Shiitake (*Lentinula edodes*):** Best grown on hardwoods like oak or maple.
     - **Oyster mushrooms (*Pleurotus ostreatus*):** Thrive on softwoods like sweetgum or poplar.
   - Allow logs to sit for 2 to 3 weeks before inoculation to neutralize natural antifungal properties. Avoid logs older than 6 months to prevent contamination by wild fungi.

2. **Drill Inoculation Holes**
   - Drill holes spaced 6 inches (15 cm) apart along the length of the log.
   - Ensure the hole diameter matches the dowels for a snug fit.

3. **Inoculate the Logs**
- Insert inoculated mushroom dowels into each hole.
- Gently tap them in using a hammer or mallet until flush with the surface.

4. **Seal with Wax**
- Melt beeswax or cheese wax in a double boiler.
- Using a paintbrush, coat each dowel and hole with wax to protect from drying out or contamination.

5. **Place the Logs in a Suitable Environment**
- Store logs in a shady, moist area near your home.
- Position logs with one end touching the ground to encourage colonization.
- Leaning them against a fence or stacking them in a crisscross pattern works well.

6. **Monitor and Wait for Fruiting**
- Check logs after rain for signs of mycelium (white strands) spreading through the wood.
- Mushrooms can take 6 to 12 months to begin fruiting, depending on the species and climate.

METHOD 2: GROWING MUSHROOMS WITH SAWDUST BLOCKS

Step Instructions

1. **Prepare the Growing Medium**
- Fill a plastic bag or bucket with moist hardwood sawdust, straw, or coffee grounds.
- Pasteurize the substrate if necessary to prevent contamination.

2. **Inoculate with Sawdust Spawn**
- Mix inoculated sawdust spawn evenly into the substrate.
- Seal the bag or loosely cover the bucket to maintain humidity.

3. **Store in a Dark, Humid Location**
- Place in a bathroom or another humid area with minimal light exposure.
- Mist daily to maintain moisture levels.

4. **Wait for Fruiting**
- Mushrooms should begin to appear within a few weeks.
- Harvest when caps are fully developed but before they begin to release spores.

Sawdust blocks inoculated with shiitake spores.

Materials Needed

→ Inoculated sawdust spawn (from a mushroom supplier)
→ Plastic bag or bucket with a lid
→ Straw, hardwood sawdust, or coffee grounds (substrate)
→ Spray bottle for misting
→ Dark, humid indoor space (such as a bathroom)

TIP: When using this method, you may experience multiple flushes of mushrooms over several months!

of the plant community. We'll dive deep into plant communities in the next section.

Dynamic accumulators are plants with deep taproots that draw minerals from the subsoil, depositing them in the upper layers for other plants to access. Examples include comfrey, yarrow, and dandelion. Nitrogen fixers, such as those in the legume family like clover, peas, and redbud, form symbiotic relationships with bacteria that convert atmospheric nitrogen into a usable form for plants. Incorporating native plants with diverse root structures, like herbaceous perennials, grasses, and shrubs, not only sequesters carbon but also supports microbial life by feeding soil organisms through their extensive root systems.

Minimize Disturbance

Tilling may be necessary during initial garden establishment, but it should not be part of your ongoing management. Think of your soil as a lasagna, with fresh organic matter on top and decomposing layers building underneath. Disturbing the soil can release carbon into the atmosphere, reducing its water-retention capacity and structure. Carbon, an essential component of soil organic matter, also plays a major role in soil fertility. Practices that reduce disturbance, like no-till farming and composting, can help maintain carbon storage in the soil.

Capturing carbon in the soil is often referred to as carbon sequestration. Plants absorb carbon from the air, and through their roots, they feed it to microorganisms in the soil. These organisms emit carbon as part of their metabolic processes. By keeping soil intact and preventing disturbance, we can retain that carbon and use it to improve soil health. According to Rattan Lal, a soil expert at Ohio State University, carbon stored in soil aggregates can remain stable for thousands of years, making it a long-term tool in fighting climate change.

When you're building soil, the less you disturb it, the better.

LEAVE THE LEAVES!

Leaves make excellent mulch and habitat.

Leaves are not a nuisance, but rather they are an incredible, free resource you can put to work in your landscape to benefit both your plants and wildlife. By keeping them on-site and using them as a mulch in your planting beds and garden, you'll add organic matter to your soil, protect roots from frost, and help retain moisture. Leaf litter also provides excellent winter cover for beneficial insects and other creatures. You can rake leaves directly onto your beds or pile them to store for mulching later or for use in composting.

Beyond the ecological benefits of keeping leaves on-site, putting them to use in your landscape has economic perks as well. By adding more organic matter and moisture to your soil over time, you'll reduce the need for irrigation and to buy fertilizer, and you'll save money on buying bags to collect them in every year. You'll decrease your carbon footprint, too!

If you're not planning to mulch your beds with leaves immediately, create a designated leaf pile or leaf cage in your landscape where they can break down further. You can add leaves to your compost pile, too!

Leaves make ideal habitats for all kinds of critters and add invaluable organic material to planting areas.

NATURAL ALTERNATIVES TO CHEMICALS IN LANDSCAPING

In the previous chapter, we explored the harmful effects of synthetic herbicides, pesticides, and fertilizers on our environment. Rather than relying on these chemicals, homeowners can adopt more sustainable practices to maintain healthy landscapes.

- **Enhance biodiversity.** Increasing plant variety in your yard makes it more resilient against pests and diseases, reducing the need for chemical interventions.
- **Prioritize native plants.** Native species have built-in defenses against local pests and diseases, requiring fewer inputs to thrive.
- **Embrace natural beauty.** A few blemishes on leaves are a normal part of nature; not every imperfection needs to be treated.
- **Cultivate healthy soil.** Soils rich in microorganisms improve fertility, increase water retention, and support beneficial plants and insects, reducing the need for fertilizers, irrigation, and pesticides.
- **Welcome beneficial insects.** Many insects play a crucial role in controlling pest populations naturally, making them valuable allies in a regenerative landscape.
- **Recognize plants as part of the ecosystem.** While it may be disappointing to see leaves nibbled by insects, many plants serve as essential hosts for pollinators and other beneficial creatures. Allowing caterpillars to feed can ultimately strengthen the ecosystem, as most plants recover quickly.
- **Grow more, mow less.** Converting sections of your lawn into gardens for food production can be both rewarding and practical. Raised beds and edible landscapes reduce the need for mowing, minimize soil compaction, and help lower carbon emissions.

Growing dense plantings keeps the soil moist, and provides ample habitat for pollinators, songbirds, and other wildlife. Increasing plant variety in your yard makes it more resilient against pests and diseases, reducing the need for chemical interventions.

SOIL AMENDMENTS

Organic soil amendments are used to improve soil structure, fertility, and microbial activity by adding decomposed plant or animal material. Common amendments include compost, manure, cover crops, and biochar, which enhance water retention, nutrient availability, and soil aeration. By incorporating organic amendments, you can accelerate soil regeneration, creating healthier, more resilient soils that support plant growth while reducing the need for synthetic fertilizers.

Here are my top five soil amendments and their benefits.

Compost

Compost is decomposed organic matter. It's made up of "green" and "brown," with green being food scraps, fresh green yard debris, grass clippings, and other sources of nitrogen, and brown being dead leaves, wood chips, sawdust, straw, paper, cardboard, branches, or other sources of carbon. The magic ratio is two parts carbon to cover one part nitrogen by volume.

Over time, thermophilic or "heat-loving" bacteria eat the organic material and produce a nutrient-rich soil amendment. The pile should be between 105°F and 140°F (40°C and 60°C) to encourage the right types of bacteria. Just like fire needs oxygen to burn, your pile needs oxygen to heat up. The ideal pile size to achieve the desired heat level is three feet by three feet (91 by 91 cm). In the section below, I'll describe the different types of home compost systems and their pros and cons, so you can implement one that works with your lifestyle.

Worm Castings

Worm castings are the waste or poo from worms, usually red wigglers. They are created through vermicomposting, where the worms eat through kitchen scraps and turn it into rich waste filled with nutrients for beneficial microorganisms. Worm castings are nutrient-rich for plant health and increase soil aeration and drainage.

Mulch

We use mulching as a stopgap while plants are establishing, rather than the intended finished ground covering. I prefer triple-shredded bark to wood chips, but if you get a pile of free wood chips delivered from a tree company, you can pre-compost them before adding them to your planting areas. Do this by letting them sit in a pile for a month or two, so they can begin decomposition. Also worth noting, you should never use fresh wood chips on your

planting beds, as they will pull the nitrogen out of your soil to jump-start decomposition, and your plants will suffer. It is also worth noting that pine straw and pine bark mulch are both very acidic, which can suppress growth of the ground cover layer. This can be useful in avoiding weeds but detrimental to your plantings and establishing a living mulch.

Compost Tea

Compost tea is made by steeping a small amount of compost in aerated water, which encourages the rapid growth of beneficial microorganisms in the water. This can then be applied as a liquid fertilizer to the soil. Elaine Ingham, a soil microbiologist with Soil Foodweb Inc., argues that the additional growth and root depth generated through the use of compost tea, applied to an area the size of California for three years, would set us back to pre-industrial atmospheric carbon levels because of all the carbon that growth would sequester. You'll find an easy step-by-step project of how to make compost tea in the next chapter on page 115.

Biochar

Biochar is an incredible and ancient soil technology made through the slow and low burning of wood, called pyrolysis. This process fixes the carbon in a form that will be stable for thousands of years, keeping it from turning into atmospheric carbon and acting as an outstanding source of soil carbon. Biochar can be inoculated with beneficial microorganisms in compost tea and introduced into the soil planting medium, where it will propagate the rapid growth of the soil microbiome and dramatically increase water infiltration.

Opposite left Making compost is an excellent way to reduce your waste and turn your kitchen scraps into fertile soil.

Opposite right Worm castings are a super-rich soil amendment that can be added to potting soil or vegetable gardens. You can also use liquid worm castings as a soil spray.

Center Mulching using wood chips or leaves helps keep moisture in the soil while keeping nutrients on-site.

Above left Compost tea, made by aerating compost in water, can be sprayed over large areas and enhances the microbial life of soil.

Above right Biochar is an excellent soil amendment to help retain moisture in the soil.

THREE HOME COMPOSTING METHODS WITH PROS AND CONS

Composting can be as simple as throwing your food scraps into a pile and covering them. Still, there are some considerations you should examine when thinking of composting in an urban/suburban area. Here are three composting methods you can implement at home and the pros and cons of each. You should know what kind of attention you can give your pile (if any) and if there are any issues with wildlife (hard lesson: in my case, it was rats) that you should be aware of when building your composting system.

THE SLOW AND LOW PALLET PILE

What is it?

The "Slow and Low" method, as I like to call it, is the lazy gardener's dream. In this method, you make a designated space in your yard where you empty your food scraps, you cover them with at least two parts carbon (read: brown material such as leaves), and you let nature do the work of breaking down the food into a rich, fertile soil amendment. Many people create multiple composting bays using recycled pallets, so once a pile is finished, you can stop adding to it and let it rest while you continue composting in a second bay.

Reusing wooden pallets is a great way to frame a compost area.

How to Do It?

It starts with identifying a good spot, far enough from your house that you aren't going to smell it on a humid summer day, but close enough that you actually use it. You can easily wrangle used pallets from a local hardware store. The two main things to know to do this method effectively are:

- COVER! COVER! COVER! Always cover your food scraps with at least twice as much carbon (brown material like leaves) than the volume of waste you are disposing of.
- Keep the pile to a three feet (91 cm) high by three feet (91 cm) wide ratio, which helps keep it hot. That way, the thermophilic bacteria that eat the waste are happy, and your pile doesn't go anaerobic.

Pros of the Slow and Low Pallet Pile

This method is cheap and it doesn't require purchasing any infrastructure. By choosing this method, you let nature do the work and patiently await soil that can be used in the garden. Plus, it's easy to get started, so you can begin composting quickly.

Cons of the Slow and Low Pallet Pile

Some things to consider before you choose this method:

- It will be smelly if you don't cover your scraps well. There is a difference between a composting pile and an anaerobic, festering pile.
- If you live in a neighborhood with an HOA or where the houses are close together, this method will likely attract your neighbors' unwanted attention if you are not discreet, especially if you are inconsistent with covering food scraps, and the pile begins to smell!
- You run the risk of attracting rats. This is a big one.

Funny story (er, hard lesson?): I used to be a huge proponent of this method. It worked great when I lived on a farm and had plenty of space and no neighbors to piss off. When I used this method in my first urban homesite, I discovered it comes with a cost in the city: rats.

For a couple of years, the pile was fairly innocuous, and I would see an occasional critter, but that changed when a neighboring house was gutted and renovated. The rats that lived in the previously unoccupied house took to the woods when construction began and were quickly attracted to my compost pile. The heat the pile emits during winter and the consistent food source were huge liabilities. Plus, this type of pile is open-air, so there is nothing to stop them from moving in and burrowing their little rat babies in the warmth of the pile. Nocturnal visits to the pile from the new rat families now residing in the woods behind my house quickly turned into unwelcome guests in my own house! Attracted to a random bag of seeds I had stored in my garage downstairs (another big mistake in this whole drama!), they took up residence in my basement, where they chewed through wires, defecated in the ceiling beneath my living room (talk about smell!) and ultimately caused a massive headache that required me to rip out the ceiling, fix wiring, and seal my crawl space.

An aboveground tumbler helps contain compost to keep out rodents and reduce odor.

THE TUMBLER

What Is It?

A compost tumbler is a freestanding, sealed rotating barrel of sorts, designed to speed up the decomposition process by allowing aeration and easy mixing of organic waste. Unlike traditional compost piles, which require manual turning with a pitchfork or shovel, a tumbler can be rotated by hand or with a crank, making composting more efficient and user-friendly. They can be made of plastic or metal.

How to Do It?

This method is extremely easy. You collect your scraps and pour them into the tumbler. Some models recommend adding carbon, and others do not. My tumbler has two compartments, so you can stop adding to one side to let it sit and add to the other side in the meantime. This gives the fallow side time to break down without disrupting your flow of composting.

Pros of the Tumbler

It's easy. It's self-contained, so, depending on the model, you don't need to worry about rodents. This was, of course, a huge deciding factor for my household, and one of the reasons I went with a tumbler. I also appreciate that mine is made of metal and is sturdier than its plastic counterparts. Some other benefits are:

- **Faster composting.** Rotating the bin introduces oxygen, which helps microbes break down organic matter quickly.
- **Odor and pest control.** The enclosed system prevents rodents, flies, and odors from becoming a problem.
- **Ease of use.** Less manual labor than traditional compost piles since turning is done by simply rotating the bin.
- **Space efficiency.** Ideal for small yards, patios, or urban gardens.

Cons of the Tumbler

Depending on how much cooking you do and how much food waste you generate, they fill up fairly quickly. It would help if you considered this when choosing the right model for your lifestyle. I have a 70-gallon (265 L) tumbler, because I do a lot of cooking. Each compartment is 35 gallons (132 L). When I first started using this method, I filled both sides within six months but had not turned it enough for the fallow side to be completely decomposed, leaving me in a pickle. I decided to take the almost-finished compost out and finish it on the ground beneath the tumbler, but if it had not been as well-decomposed as it was, I would be very reluctant to empty the compartment so as not to attract rats again. Another con is that compost tumblers can get expensive quickly, especially if you choose a larger metal one. There are definitely some budget-friendly options, though, so take a peek on the internet if this is something you are considering.

Turn scraps into soil! A compost subscription service makes it easy to reduce waste—just fill your bin and they do the rest. Pros: eco-friendly, no mess. Cons: added cost and varying pickup schedules.

A COMPOST SUBSCRIPTION SERVICE

What Is It?

A subscription composting service, such as Compost Now in Atlanta, is a weekly or biweekly pickup service that collects your compost and finishes it off-site, often giving finished compost to its patrons.

How to Do It?

Usually, you are given a bucket with specific instructions about what and what not to compost. You fill the bucket (and cover it with a lid) and then depending on the volume of waste you generate, the company picks up your bucket either weekly or biweekly.

The Pros of a Subscription Compost Service

This is a fantastic option if you want to divert waste from the landfill, but are wary of composting at your home, either because you feel daunted by the prospect or you simply do not have the space (i.e., apartment dwellers). And bonus: You get access to a fertile, nutrient-rich soil amendment that is professional quality compost you can use on your site or in your houseplants without the labor of composting yourself.

The Cons of a Subscription Compost Service

The main downside is the cost. Subscription services start at about $20 USD per month and can go up from there. Also, subscription services are not available everywhere, so depending on where you live, this might not even be an option! A little internet research should answer that pretty quickly.

While these are obviously not the only three methods for home composting, they are, in my opinion, the most popular and most accessible. Ultimately, however, the responsible management of organic waste shouldn't fall entirely on the consumer's shoulders. Many municipalities are exploring curbside compost pickup, just as your trash and recycling are picked up. This can be of great economic benefit to cities that can then sell the finished compost to residents for a premium. Some cities in the United States, such as Seattle, Washington, and Boulder, Colorado, as well as several cities in Europe, have programs like this already in place. The most significant barrier to this type of program for most cities is the infrastructure cost to implement it on a broad scale, but it has been proven possible.

Pest Management Begins with Soil Health

Pest management, from an ecological perspective, is deeply rooted in the principles of permaculture, which focuses on working with nature rather than against it. Rather than relying on chemical pesticides that disrupt ecosystems, permaculture-based pest management creates balanced, self-regulating systems that naturally control pest populations.

At its core, permaculture pest management integrates biodiversity, soil health, habitat creation, and natural predator-prey relationships to reduce pest pressure without compromising the health of the environment. Instead of viewing pests as isolated problems, we can see them as indicators of imbalance—a sign that the ecosystem needs adjustment.

WHAT MAKES A PEST, A PEST?

The conventional understanding of a pest is a destructive, invasive, or harmful insect, mollusk, or plant that works to destroy a garden. Mites, beetles, flies, slugs, even certain moths and caterpillars—the list goes on. However, if regeneration starts with the foundational belief that "it's all alive; it's all intelligent; it's all connected," where does that leave us when it comes to pests? It points to intrinsic value and all life serving a purpose within an ecosystem.

ALL LIFE FORMS HAVE INTRINSIC VALUE

The permaculture ethic of "intrinsic value" recognizes that all living beings, ecosystems, and natural elements have inherent worth, independent of their usefulness to humans. This ethic extends beyond sustainability, encouraging a deep respect for all forms of life and acknowledging that nature is not just a resource to be extracted from but a living system that deserves care and protection in its own right. In order to address a pest in a garden, start from a place of curiosity about its intrinsic value, which can lead to a more connected understanding of its role in our landscape.

Let's look at a few examples of plants and insects that are often considered pests in the garden.

Opposite, top Native to North America, the polyphemus moth (*Antheraea polyphemus*) is a vital part of the food web. Its caterpillars feed on the leaves of native trees like oak, maple, and willow, while birds, bats, and parasitic wasps rely on both the caterpillars and adult moths as a food source. Though the adults live only a few days and don't eat, their presence signals a balanced, biodiverse ecosystem.

Opposite, bottom If something isn't eating your garden, it isn't part of the food web. The gulf fritillary (*Agraulis vanillae*) caterpillars feed exclusively on passionflower vines (*Passiflora* spp.), making this plant essential to its lifecycle. In turn, the butterfly provides food for birds, spiders, and predatory insects, while adults pollinate flowers as they feed on nectar.

Poison Ivy (*Toxicodendron radicans*)

We often have clients concerned about the growth of poison ivy in their yard. Most people are allergic to poison ivy, and it can sneak up and grow in a variety of places and conditions. However, let's look at the value of poison ivy and its purpose within the ecosystem. Poison ivy grows best as a wood's edge plant, and it protects the wild forest. It thrives along the perimeter of new developments that encroach upon wild forests. It also thrives in suburban settings where a manicured lawn meets the edge of a wooded area. Understanding the intrinsic value of this protector plant helps reframe how we think about its place in a garden. Beyond where it grows, the seeds and berries of poison ivy are important winter food sources for threatened woodland songbirds, such as woodpeckers (*Picoides* spp.), northern flickers (*Colaptes auratus*), and yellow-bellied sapsuckers (*Sphyrapicus varius*).

Fire Ants (*Solenopsis* spp.)

Fire ants build ant hills in inconvenient spots and sting if their nests are disturbed. Yet, it's the ant hill that points to the intrinsic value of a fire ant within an ecosystem. As they build their hill, they tunnel into the earth and push soil up to the surface. They break up even the most compacted, clay-heavy soils, making them more porous, whereby more water infiltrates into the ground. As a result, plants can draw water from the soil during periods of drought and the soil becomes more nutrient rich.

WHAT IS A PEST'S PLACE WITHIN THE FOOD WEB?

This is an important question because the reality is, by the time conventional solutions address a pest problem, they're looking downstream and treating symptoms instead of looking at the issue holistically. Instead, let's consider the food web—what eats certain pests, and what do certain pests eat? Here are some examples.

Aphids (*Aphis fabae* and Others in the Superfamily Aphidoidea)

Aphids are tiny sap-sucking insects that can affect a wide variety of plants, and they reproduce rapidly. They feed off nutrient-rich plants and large numbers can cause damage. However, natural predators, such as ladybugs (*Coccinella septempunctata* and others) and lacewings (*Chrysoperla carnea*) can help control populations. Hoverflies (*Syrphus ribesii*), which are crucial pollinators that hover over flowers to drink nectar, lay their eggs close to aphids because once they hatch, the larvae feed on aphids. Aphids supply food to beneficial larvae and help keep balance within the food web.

Black Widow Spiders (*Latrodectus* spp.)

We recently had a client with a growing black widow population in her backyard. Black widow spiders are venomous arachnids known for their glossy black bodies, red hourglass markings, and potent neurotoxic venom, though they are generally shy and bite only in self-defense. Needless to say, the client was perturbed—not wanting to spray, but also not wanting to worry about her young daughter every time she was outside in the yard. After some quick research—asking the question "What eats black widows?"—I learned that a wasp called the iridescent blue mud dauber is the primary predator of the black widow. By introducing this beautiful species, the client could address the overpopulation of black widows to become more balanced within the food web (see sidebar page 102).

HOW DO YOU INTRODUCE A SPECIES LIKE THE IRIDESCENT BLUE MUD DAUBER?

The iridescent blue mud dauber (*Chalybion californicum*) is a solitary, non-aggressive wasp known for its striking metallic blue color and natural pest control abilities, particularly against black widow spiders and other arachnids. It's found across North America. Encouraging these beneficial wasps in your landscape can help maintain ecological balance without the need for chemical interventions.

Steps to Attract and Support Blue Mud Daubers

Provide Nesting Sites

Unlike social wasps, mud daubers are solitary nesters that build mud tube nests in protected areas. To encourage their presence:

- Leave undisturbed structures such as old sheds, eaves, and rafters where they can build their mud nests.
- Consider providing artificial nesting sites such as wooden boards or sheltered wall spaces.

Ensure a Reliable Mud Source

As their name suggests, mud daubers rely on moist soil or mud to construct their nests. To support them:

- Maintain a damp, muddy area near their nesting sites, especially in dry climates.
- Use drip irrigation or shallow water trays filled with soil to create a mud source.

Encourage a Steady Spider Population

Mud daubers primarily hunt spiders (including black widows, orb weavers, and jumping spiders) to feed their larvae. A healthy, pesticide-free environment with natural insect diversity ensures that they have ample food.

- Allow native plant diversity to attract insect prey, which in turn supports a stable spider population.
- Avoid broad-spectrum insecticides, which can disrupt the predator-prey balance and harm beneficial species.

The iridescent blue mud dauber (*Chalybion californicum*) is a solitary, non-aggressive wasp known for its striking metallic blue color and natural pest control abilities, particularly against black widow spiders and other arachnids.

Plant Pollinator-Friendly Flowers

Although adult mud daubers primarily hunt spiders, they also drink nectar from flowers. Planting nectar-rich flowers will provide them with energy and encourage their presence. Some good options include:

- Milkweed (*Asclepias* spp.)
- Goldenrod (*Solidago* spp.)
- Mountain mint (*Pycnanthemum* spp.)
- Coneflowers (*Echinacea* spp.)

Avoid Disrupting Existing Nests

If you find an established mud dauber nest, avoid removing or disturbing it. Unlike paper wasps or hornets, mud daubers are not aggressive and will not defend their nests. They often reuse old nests, so leaving them in place encourages their continued presence.

Caterpillars

Notoriously very hungry, caterpillars eat plants and flowers, leaving them perforated with holes or even depleted of nutrients. But not only do caterpillars become butterflies or moths, which are critical pollinators, they also cycle nutrients and energy through the ecosystem for other plants and living things. Plus, butterflies and moths strategically lay their eggs where sustenance is available for their larvae once hatched. For example, the gulf fritillary butterfly (*Agraulis vanillae*) lays her eggs along the passionflower's stem or underneath its leaf because the passionflower is a food source for the caterpillar. A monarch butterfly (*Danaus plexippus*) will only lay her eggs on the milkweed (*Asclepias* spp.) plant, which eventually feeds the resulting caterpillars.

Ultimately, remember that a regenerative landscape relies on diversity, and it thrives if all living things are interdependent and connected.

An ecosystem cannot thrive if it serves just one purpose or if it prioritizes certain plants and living organisms over another. When thinking about pests, ask a few questions first:

- What is the intrinsic value of this life form in my garden?
- Is it performing a function from which something else benefits?
- Is it a food source for another creature?
- What can I introduce that eats this pest?
- What can I plant to feed its predators, in order to bring more balance?

Working with Nature

A circular economy in the garden is about working with natural cycles to reduce waste, regenerate soil, and create self-sustaining systems. By composting organic materials, using natural mulches, and designing for biodiversity, we can build soil health while minimizing reliance on external inputs. This approach not only benefits the environment but also makes landscapes more resilient and productive over time. By keeping resources in use and supporting natural regeneration, we shift from extracting nutrients to cycling them, strengthening the foundation of our ecosystems. In the next chapter, we'll explore specific soil-building techniques that will help you enhance soil fertility and structure in your permaculture landscape. We'll take a closer look at practical techniques to build healthy soil in your own yard. Whether it's using hügelkultur beds, lasagna gardening, or compost tea, these methods will help you create a nutrient-rich, moisture-retaining soil ecosystem that supports abundant plant growth.

While these chickens are taking a break in their backyard coop, they also free range throughout the garden. Doing so helps minimize the life cycle of certain pests and introduces nutrients to the soil.

CHAPTER 6

SOIL TECHNIQUES

Improving soil quality takes patience, but it's one of the most valuable investments you can make in your landscape. Healthy soil forms the backbone of a thriving ecosystem, supporting plant life above ground and a complex web of organisms below. When soil is rich in organic matter and biodiversity, it becomes more porous and nutrient-dense, making it easier for plant roots to grow and for water to infiltrate.

Because traditional landscaping often treats yard debris as waste, focusing on its removal rather than its potential, it leads to unnecessary resource consumption—fuel for mowers and blowers, transportation to disposal sites, and the production of waste collection bags, to name a few. Clearing away organic material like leaves and trimmings also strips the landscape of valuable resources that could otherwise enhance soil health and help retain water and carbon. Additionally, frequent mowing and leaf blowing contribute to carbon emissions, worsening climate change and reducing the landscape's ability to withstand extreme weather conditions.

A better approach is to see organic debris as a resource that can be reintegrated into the landscape to build soil. Some effective ways to do this include:

- **Using small livestock in closed-loop systems (where permitted).** Animals like chickens and rabbits produce nutrient-rich manure for composting while also naturally grazing and fertilizing as they move through the landscape.
- **Composting yard and kitchen waste on-site.** Properly maintained compost piles (see page 96) generate heat, accelerate decomposition, and eliminate pathogens, producing nutrient-dense soil amendments.
- **Retaining leaves for mulch.** Instead of removing leaves, use them as a natural mulch to help retain soil moisture and create habitat for beneficial insects and microbes.
- **Suppressing weeds with sheet mulching.** Layering used cardboard over garden beds helps control weeds while maintaining soil moisture.

OBSERVE AND INTERACT: ASSESS AND IMPROVE YOUR SOIL

Healthy soil is the key to a thriving garden. Understanding your soil type and how it interacts with water and plants will help you make better decisions when planning your landscape.

Step 1: Identify Your Soil Type

Take a handful of damp soil and squeeze it in your hand:

- **Clay soil:** Feels sticky and holds its shape. It retains moisture well but can become compacted and drain poorly.
- **Sandy soil:** Feels gritty and falls apart easily. It drains quickly but struggles to hold nutrients.
- **Loamy soil:** Feels soft, holds together slightly, and is rich in organic matter. This is ideal for most plants.

Step 2: Look for Problem Areas

- **Exposed soil:** Bare ground is prone to erosion and nutrient loss. Cover it with mulch or ground cover plants.
- **Standing water:** If water sits on the surface after rain, the soil may be compacted and need aeration or organic matter.
- **Dry, cracked areas:** These indicate poor water retention, and can be improved with compost and mulch.

Step 3: Improve Soil Health

- **For clay soil:** Add organic matter like compost to break up compaction and improve drainage.
- **For sandy soil:** Use mulch and compost to increase water retention.
- **For low organic matter:** Apply a layer of mulch each season and avoid removing fallen leaves to let them decompose naturally.

By identifying and improving soil conditions, you can create a more resilient and self-sustaining garden.

- **Cover cropping in vegetable gardens.** Planting cover crops between growing seasons replenishes essential nutrients, particularly nitrogen, and provides organic material for "green manure."

In the next section, we'll explore a few of these simple yet effective techniques in greater detail. Implementing these strategies can help create a vibrant, resilient soil ecosystem that supports diverse plant life, improves water retention, and builds long-term fertility—all while reducing waste and embracing a more sustainable gardening approach.

Hügelkultur: Building Soil with *Hügelkultur* Beds

Hugel-what?

Austrian farmer Sepp Holzer popularized the technique known as *hügelkultur*, which translates to "mounded beds." *Hügelkultur* beds are an excellent way to create raised garden spaces by stacking layers of decomposing logs, branches, leaves, and other organic material before covering them with soil. Over time, these beds naturally retain moisture, encourage beneficial soil microbes, and steadily release nutrients as the organic material breaks down. *Hügelkultur* combines water conservation, organic waste management, and soil regeneration in one simple method.

While this technique offers many advantages, properly constructing a *hügelkultur* bed can be a bit more complex than it seems—especially for beginners. The first challenge is sourcing suitable hardwood logs at the right stage of decomposition. If the logs are too large and freshly cut, they will take longer to break down and won't hold moisture as effectively. Those with access to forested areas may find this easier, but for others, locating the right materials in sufficient quantities can be tricky.

Another potential hurdle is the labor required to prepare the site. To build a proper *hügelkultur* bed, you need to remove existing grass and vegetation and dig

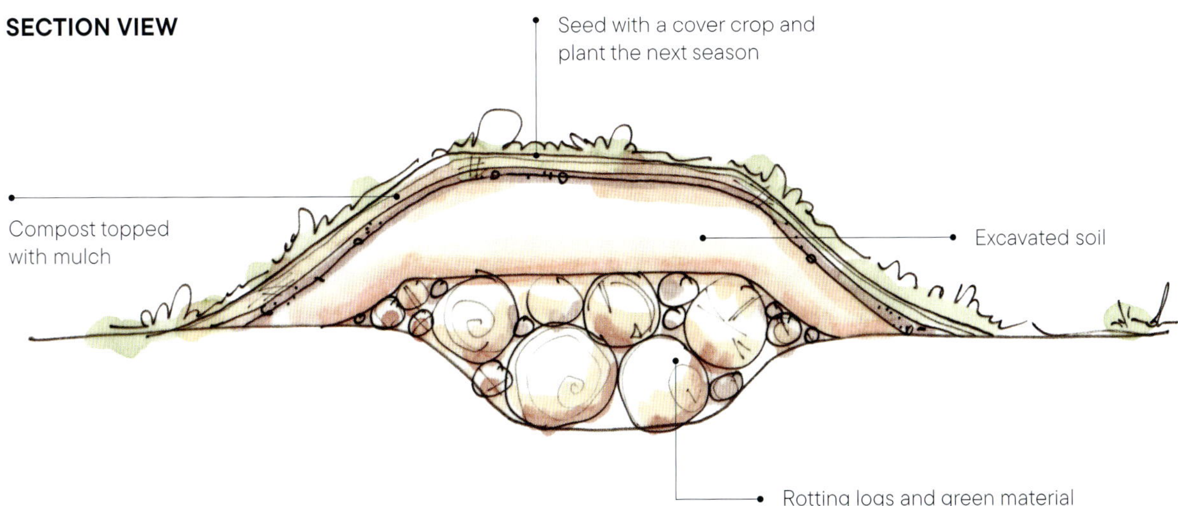

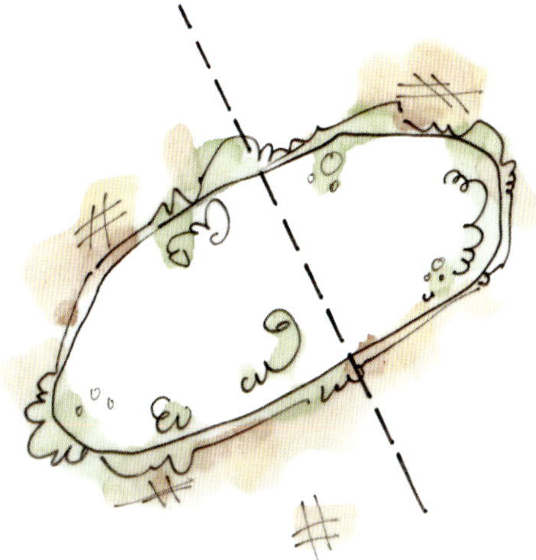

Cross section of a *hügelkultur* bed.

roughly two feet into the soil to establish the foundation. Assembling and layering materials correctly also takes time and effort, making it a demanding project for those new to gardening. However, for those willing to put in the work, the long-term benefits—improved soil fertility, better water retention, and reduced maintenance—make *hügelkultur* a valuable and rewarding addition to any garden.

One of the greatest benefits of *hügelkultur* beds is their ability to hold moisture like a sponge. The decomposing wood at the core absorbs water during rainy periods and releases it slowly, reducing the need for frequent watering. This makes *hügelkultur* an excellent strategy for drought-prone areas or sites with poor soil conditions. Over time, the materials break down, improving soil structure, increasing microbial activity, and providing plants with a slow-release nutrient source. If you're ready to try this regenerative gardening method, here's how to build your own *hügelkultur* bed.

MAKING A *HÜGELKULTUR* BED

STEP 1: DIGGING THE BASE

Begin by digging a trench along the natural contour of the land, about 2 feet (61 cm) deep and 2 feet (61 cm) wide. Set the excavated soil aside, as you will use it later to cover the *hügelkultur* bed. Digging along contour is essential for capturing and slowing down water movement, ensuring deep infiltration and preventing runoff. This design helps maximize water retention and reduces soil erosion, making it particularly useful in sloped landscapes.

STEP 2: BACKFILLING WITH WOODY MATERIAL

At the base of the trench, lay down a foundation of rotting logs, branches, and other woody debris. This material serves as the spongy core of the *hügelkultur* bed, soaking up water and gradually releasing moisture back into the soil.

- Use well-rotted logs when possible, as they will decompose faster and be more effective in moisture retention.
- Hardwoods like oak or maple will break down more slowly, providing a long-term water reservoir.
- Avoid allelopathic wood like black walnut, which release chemicals that can inhibit plant growth.

STEP 3: ADDING NITROGEN-RICH MATERIALS

Next, layer in nitrogen-rich organic matter, such as green yard waste, kitchen scraps, grass clippings, manure, or unfinished compost. Since woody materials are carbon-heavy and can temporarily tie up nitrogen as they decompose, adding nitrogen-rich inputs balances the process, ensuring plants have enough available nutrients.

Don't be afraid to mound this layer above the existing ground level—*hügel* beds are designed to be raised. The higher you build, the more organic material you can incorporate, leading to a longer-lasting nutrient cycle.

STEP 4: COVERING THE BASE LAYERS

Now, use the excavated soil from Step 1 to cover the entire mound, creating a protective top layer. This anchors the structure, prevents nutrient loss, and begins the process of integrating the decomposing materials into the surrounding landscape.

STEP 5: FINISHING TOUCHES

To jump-start soil fertility, spread a layer of finished compost over the surface. Then, overseed the mound with a cover crop mix, such as clover, buckwheat, or winter rye, to stabilize the bed and introduce organic matter as the cover crop decomposes. Finally, cover the entire bed with wheat straw or another mulch to retain moisture, suppress weeds, and protect the soil structure.

Water the seeds in thoroughly, and that's it! Over time, the wood will break down, improving soil health and creating a thriving, moisture-efficient planting space that requires minimal maintenance.

By using natural processes to regenerate soil, *hügelkultur* beds create a thriving, self-sustaining ecosystem that supports healthy plant growth, reduces labor, and enhances biodiversity. Whether you're dealing with poor soil, water scarcity, or compacted ground, *hügelkultur* offers a low-maintenance, long-term solution for building rich, productive garden beds!

Lasagna Gardening

The phrase "Lasagna Gardening" was coined by Patricia Lanza, who popularized the no-dig, sheet-mulching gardening method in her book *Lasagna Gardening: A New Layering System for Bountiful Gardens* (1998). The method involves layering organic materials like newspaper, cardboard, compost, and mulch to create rich, fertile soil without tilling. There are three techniques here that can each be employed independently, but I've found that they work really well together to rapidly build soil.

The combination of double digging, sheet mulching, and chop and drop weeding is extremely effective. In fact, when I tested it in my own backyard, where the soil was so depleted and compacted that I was unable to dig into it with a shovel, I was able to rapidly build a layer of topsoil nearly twelve inches (30 cm) deep in just two growing seasons. Now, I can reach my hand into the loamy soil, nearly up to my elbow!

DOUBLE DIGGING

Double digging involves turning the topsoil over and burying the ground cover layer with subsoil, allowing it to decompose and enrich the soil. This method, often combined with sheet mulching, is highly effective for transforming compacted or grassy areas into fertile planting zones. Double digging increases soil aeration, improves root penetration, and enhances nutrient availability.

This practice is a labor-intensive but highly effective soil preparation technique. By turning over the topsoil and burying organic material in the subsoil, the decomposing plant matter feeds soil microbes, enhances nutrient availability, and promotes deep root growth.

Double digging is especially beneficial for establishing no-till gardens in heavily compacted, nutrient-depleted, or grassy areas. When combined with sheet mulching, it creates rich, well-structured soil that supports long-term plant health.

HOW TO PREVENT GRASS FROM GROWING BACK WHEN DOUBLE DIGGING

When double digging, grass and weeds can regrow if not properly managed. Simply turning the soil isn't always enough, as grass roots and rhizomes can persist and sprout again. Here are some effective strategies to ensure grass doesn't return after double digging.

Step 1: Turn the Grass Upside Down (Sod Inversion)

- When digging the trench, flip the grass upside down so that the roots are uprooted and then completely buried under the soil.
- This blocks sunlight, preventing the grass from photosynthesizing and regrowing.
- Over time, the buried grass decomposes, adding organic matter to the soil.

Step 2: Smother with Sheet Mulching

- After double digging, lay down a thick layer of cardboard or newspaper over the soil to block grass and weed regrowth.
- Cover with 3 to 6 inches (8 to 15 cm) of mulch (straw, wood chips, or leaves) to hold moisture and suppress any remaining roots.
- If planting immediately, add a layer of compost on top so seeds or transplants have access to soil.

HOW TO DOUBLE DIG: A STEP-BY-STEP GUIDE

STEP 1: GATHER YOUR TOOLS AND MATERIALS

- **Tools:** A shovel, digging fork, wheelbarrow, and a rake.
- **Materials:** Compost, mulch, and nitrogen-rich organic matter (such as grass clippings, food scraps, or green manure).

STEP 2: MARK AND PREPARE THE AREA

- Choose the area you want to improve and mark off a section 2 to 3 feet (61 to 91 cm) wide.
- If there is existing vegetation (grass or weeds), cut it back to make your digging easier.

STEP 3: DIG THE FIRST TRENCH

- Using a shovel, dig a trench 6 inches (15 cm) deep and 6 inches (15 cm) wide along the length of your marked section.
- Separate the green upper layers from the subsoil. Place the excavated soil in a wheelbarrow or pile it nearby, as it will be used later. Place the green material to the side.

STEP 4: LOOSEN THE SUBSOIL

- Use a digging fork to loosen the subsoil (the compacted layer beneath the topsoil) another 6 inches (15 cm) deep, but do NOT remove it.
- This extra depth improves water infiltration, root penetration, and aeration.

STEP 5: ADD ORGANIC MATERIAL TO THE TRENCH

- Lay down the layer of green organic matter you just excavated at the bottom of the trench.
- This buried material gradually decomposes, enriching the soil with nutrients over time.

STEP 6: DIG THE NEXT TRENCH AND TRANSFER SOIL

- Move one step and dig a second trench adjacent to the first one.
- Take the excavated topsoil from this new trench and place it into the first trench, covering the organic matter.
- Continue this process of trenching and soil transfer until the entire bed is double-dug.

STEP 7: FINISH BY ADDING MULCH AND COMPOST

- Once you have reached the end of your plot, fill the last trench with the soil you originally set aside from the first trench.
- Spread a thick layer of compost over the entire area, followed by mulch (straw, leaves, or wood chips) to retain moisture and protect the newly aerated soil.

Meadows keep soil covered year-round, preventing erosion and locking in moisture. They can be established after sheet mulching or using the chop and drop method.

SHEET MULCHING

Sheet mulching is a technique that uses cardboard topped with compost or mulch to suppress weeds and enhance soil fertility. By laying cardboard over the soil, you create a barrier that encourages the green layer beneath to decompose, adding valuable organic matter. This method is especially effective for converting grassy lawns into productive garden beds, reducing labor, and minimizing the need for chemical weed control.

Remember: If you're transitioning a grassy area to a garden bed, or working with depleted soils, I recommend double digging the area first. If you want to build soil over several seasons—maybe because your soil is so compacted you can't work it yet, or because you're trying to decide what to do with an area of your yard—I recommend combining sheet mulching with the chop and drop method, which we'll dive into next.

CHOP AND DROP: ACCELERATING SOIL BUILDING WITH COVER CROPS

At the end of the growing season, if you want to boost soil fertility and organic matter, you can use a technique known as chop and drop—a simple but effective method that allows cover crops to decompose in place, enriching the soil naturally. Instead of removing plant material or tilling the soil, you cut down your cover crop at ground level and leave it as mulch, allowing it to break down over time and feed soil microbes.

This method mimics natural forest ecosystems, where organic material continuously falls to the ground, slowly decomposing and cycling nutrients back into the soil. By leaving plant matter in place, you create a protective mulch layer that enhances soil moisture retention, suppresses weeds, and prevents erosion.

Benefits of Chop and Drop for Soil Health

- **Boosts soil organic matter.** Decomposing plant material feeds soil microbes and improves nutrient cycling.
- **Eliminates the need for tilling.** Avoids soil disruption, preserving mycorrhizal networks and soil structure.
- **Retains moisture.** Acts as a mulch layer that reduces evaporation, keeping soil hydrated.
- **Suppresses weeds.** Smothers weed growth and reduces competition for future plantings.
- **Reduces labor.** No need to remove plant material or haul away biomass—everything stays on-site to decompose naturally.

By using the chop and drop method, you can build soil health year after year, reduce inputs, and create a thriving, self-sustaining ecosystem. Whether you replant immediately or let the soil rest, this low-maintenance, high-reward technique is a key part of regenerative soil management.

HOW TO LAY CARDBOARD FOR SHEET MULCHING

Overlap the edges of your cardboard so weeds don't grow up around the edges.

Wet the cardboard down so it is flat and molds to the ground beneath.

Cover the cardboard with compost or mulch.

STEP 1: MOW OR CUT DOWN EXISTING VEGETATION

No need to till; just flatten the area as much as possible.

STEP 2: WATER THE SOIL THOROUGHLY

Moist soil speeds up decomposition and helps the cardboard settle.

STEP 3: LAY CARDBOARD OVER THE SOIL

- Use uncoated, brown cardboard (avoid glossy or plastic-lined materials).
- Overlap the edges by at least 6 inches (15 cm) to block weed growth.
- Remove any tape or staples before laying the cardboard.

STEP 4: WATER THE CARDBOARD WELL

This helps it stick to the ground and break down faster.

STEP 5: ADD LAYERS OF ORGANIC MATERIAL ON TOP

Compost, mulch, or soil will weigh it down and feed the soil as it decomposes.

HOW TO IMPLEMENT CHOP AND DROP

STEP 1: CUT COVER CROPS AT GROUND LEVEL

If you have an existing cover crop (such as clover, rye, vetch, or buckwheat), begin by:

- Cutting the plants at ground level, leaving the roots intact to decompose and improve the soil structure.
- Leaving the chopped biomass in place as the first nitrogen-rich layer of organic matter. This initial layer jump-starts decomposition and adds green material (nitrogen) to the soil while maintaining soil life and moisture.
- Use shears, a scythe, or a mower to chop cover crops down to the ground.
- Leave the roots intact—they will continue to decompose underground, creating natural air pockets and improving soil structure.

STEP 2: LEAVE THE BIOMASS AS MULCH

Scatter the chopped plant material evenly across the soil surface to act as an organic mulch. This layer will protect the soil from temperature fluctuations, retain moisture, and provide slow-release nutrients as it breaks down.

STEP 3: SHEET MULCH OVER THE CHOPPED BIOMASS

To continue to build soil, you can sheet mulch over the chopped biomass using cardboard, compost, or straw, effectively smothering weeds while allowing organic matter to decompose beneath the mulch. You can continue to repeat this process of sheet mulching → cover cropping → chopping and dropping → sheet mulching over several growing seasons if you're starting with very depleted or compacted soils, as I was.

- Cover the chopped cover crop layer with cardboard, newspaper, or brown kraft paper to suppress weeds and block sunlight.
- Overlap pieces by at least 6 inches (15 cm) to prevent weeds from poking through.
- Lightly wet the material to help it break down faster.
- This layer smothers weeds and grass while allowing decomposition to happen underneath, creating a rich composting zone in place.

If you're going to continue with sheet mulching again, mimic the layering process of a compost pile to create a nutrient-dense growing medium. Alternate between:

- Nitrogen-Rich (Green) Layers: kitchen scraps, coffee grounds, grass clippings, fresh manure, or garden trimmings.
- Carbon-Rich (Brown) Layers: leaves, straw, wood chips, dried plant material, shredded paper, or sawdust.

Each layer should be about 1 to 2 inches (3 to 5 cm) thick, with several alternating layers to promote microbial breakdown and soil enrichment.

STEP 4: RE-SEED WITH ANOTHER COVER CROP

To keep building soil for the next season, you can seed directly into the chopped material. This works well for crops like clover (*Trifolium* spp.), vetch (*Vicia* spp.), or winter rye (*Secale cereale*).

- Spread a thick layer of compost on top to create a planting surface.
- Add a final mulch layer (such as straw, wood chips, or leaf litter) to lock in moisture and prevent erosion.
- Water the entire bed thoroughly to activate decomposition.

If possible, it's best to allow the bed to break down over a few months (especially if it is built in the fall or winter) before planting in your freshly made soil for a richer, more stable soil environment.

COMPOST TEA

Compost tea is a liquid soil amendment made by steeping compost in aerated water, promoting the rapid growth of beneficial microorganisms. When applied to soil or plants, it enhances microbial diversity, improves nutrient cycling, and strengthens plant resilience against pests and diseases. By introducing active soil life, compost tea helps improve soil structure, increase water retention, and support root development. This natural fertilizer not only boosts plant health and productivity but also contributes to long-term soil regeneration, making it a valuable tool in sustainable and regenerative agriculture.

Benefits of Compost Tea

Using compost tea instead of, or in addition to, solid compost has several advantages, particularly in terms of efficiency, application, and microbial activity. Here are the key benefits of compost tea over traditional compost.

Faster Nutrient Absorption

Compost tea delivers nutrients in a liquid form, making them immediately available to plants through foliar sprays or root drenching, whereas solid compost takes longer to break down and release nutrients.

Increased Microbial Diversity

Beneficial microorganisms, such as bacteria, fungi, protozoa, and nematodes, multiply rapidly after aerated compost tea applications, helping restore soil biology more effectively than compost alone.

Better Soil Penetration

As a liquid, compost tea can seep into compacted or depleted soils more efficiently than solid compost, improving soil structure and root access to nutrients.

Disease Suppression

The beneficial microbes in compost tea help outcompete harmful pathogens, protecting plants from diseases and reducing the need for chemical fungicides.

Cost-Effective and Resource-Efficient

A small amount of compost can be stretched further by brewing it into tea, making it an efficient way to distribute nutrients and beneficial microbes over a larger area.

Foliar Application Benefits

Unlike solid compost, compost tea can be sprayed directly onto plant leaves, enhancing nutrient uptake and acting as a protective barrier against pathogens.

Improved Water Retention and Drought Resistance

The microbial life introduced by compost tea enhances soil aggregation, leading to better water infiltration and retention, which supports plant growth in dry conditions.

Reduced Waste and Leaching

Liquid application minimizes nutrient runoff and leaching compared to solid compost, making compost tea a more environmentally friendly choice in areas prone to erosion or nutrient loss.

With a solid foundation of water management and healthy soil, we can now turn our attention to the plants that will bring our landscapes to life. The conventional approach to gardening often prioritizes aesthetics over ecological function, favoring non-native species and monocultures that require constant inputs. Section 3 will introduce a different approach—one that embraces biodiversity, supports pollinators, and provides food, medicine, and habitat.

HOW TO MAKE COMPOST TEA

The first step in making compost tea is to use high-quality compost. Look for compost with a dark brown or black color, similar to the rich soil found on a forest floor. If it's too light in color, it may have lost some of its nutrients from sitting too long. Texture is also key—I prefer a fine, well-decomposed compost that has been properly sifted. Smell matters just as much; good compost should have an earthy scent, not an ammonia-like odor. If it smells like ammonia, it may not have decomposed fully or was not processed correctly.

Once you've confirmed your compost is high quality, follow a simple ratio of one handful of compost per gallon (4 L) of water. For a 5-gallon (19 L) batch, place five handfuls of compost into a paint strainer bag (or pantyhose works, too!) and secure it to the side of a 5-gallon (19 L) bucket. Make sure the bag is slightly elevated, so it floats in the water rather than resting on the bottom.

Next, slowly pour water over the bag, filling the bucket to the top. Rainwater or another clean, chemical-free water source is ideal since municipal tap water often contains chlorine, chloramines, and other additives that can harm beneficial microbes. If tap water is your only option, consider using an RV filter on your hose to reduce these chemicals before brewing.

You can use compost tea as a drench to help give struggling street trees a boost.

One 5-gallon (19 L) batch of compost tea can cover up to 1 acre (0.4 ha)!

Once the bucket is filled with water, place two aquarium bubbler stones inside to keep the water aerated during the brewing process. Let the compost tea brew at room temperature for 48 to 72 hours, with a minimum of 24 hours needed for proper activation. After brewing, use the compost tea as soon as possible—once aeration stops, the beneficial microbes begin to decline, so it's best to apply it right away for maximum effectiveness.

In spring, compost tea can be applied both as a foliar spray and as a soil drench to nourish plant roots directly. When using it as a foliar spray, it's best to apply it in the early morning, allowing enough time for the leaves to dry. This helps prevent excess moisture, which can encourage fungal growth, and reduces the risk of foliage burn from intense midday sun.

All you need to make compost tea at home is compost, a 5-gallon (19 L) bucket, chemical-free water, a paint strainer bag, and aquarium bubblers for aeration.

Honeybees sip nectar from a sunflower.

Section 3

Plants

Pillar #3: Preserve and Restore Biodiversity by Building Plant Communities

Plants are the architects of life. They stabilize soil, filter water, regulate climate, and provide food and shelter for countless species—including us. Yet, conventional landscaping treats plants as decoration rather than as the essential, interconnected organisms they are. Lawns dominate urban and suburban spaces, requiring endless inputs of water, fertilizer, and pesticides to survive. Non-native ornamentals replace diverse plant communities, offering little to no benefit for pollinators and wildlife. The result? Landscapes that demand constant maintenance while contributing to biodiversity loss and ecosystem collapse.

But there's another way.

In this section, we'll explore how to design plant communities that work in harmony with nature, rather than against it. You'll learn how to build biodiversity while feeding yourself and wildlife, and to select plants that regenerate the soil, attract pollinators, and thrive without synthetic additions. From polyculture lawns to edible landscapes, this section will give you the tools to create a lush, abundant garden that supports both ecological health and human needs.

By shifting from conventional landscaping to regenerative planting, we not only restore habitat and increase resilience, but we also reclaim our role as stewards of the land. Every native tree, every pollinator-friendly flower, every edible perennial we plant is a step toward a healthier, more self-sustaining ecosystem. The question is no longer just what we want to plant—but how can we restore what has been lost and grow a future filled with abundance.

"Don't plant plants! Plant ecosystems!"

DAVE JACKE

Many suburban landscapes showcase large expanses of lawn using high-maintenance plants, while removing any ground cover layers in planting beds. This introduces a number of issues we'll discuss in this chapter.

CHAPTER 7

THE PROBLEM WITH CONVENTIONAL PLANT CHOICES

Resilient plant communities nourish people, support wildlife, and regenerate the soil, yet conventional landscaping often prioritizes aesthetics over ecological function. This approach favors uniformity, ornamental species, and high-maintenance designs that require excessive water, fertilizer, and pesticides. The unintended consequence is fragile landscapes that struggle against pests, deplete soil health, and provide little value to local ecosystems. By observing the vegetation already growing on a site, we can identify natural planting layers and ecological patterns that guide sustainable design.

Understanding local ecotypes helps us choose plants that thrive in the land's unique conditions, ensuring long-term resilience. Selecting the right balance of native and adapted species while managing aggressive plants fosters biodiversity without causing ecological harm. Thoughtfully assembling plant communities creates productive, self-sustaining systems that provide food, habitat, and soil enrichment, building a regenerative and thriving landscape.

How Conventional Plant Choices Undermine Resilience

The widespread preference for ornamental plants, chosen primarily for their aesthetic appeal and availability, often comes at the cost of environmental resilience. While these plants may look beautiful, they present several ecological challenges that weaken the health of landscapes and the broader environment.

Many ornamental plants are highly cultivated, bred for specific visual traits rather than for adaptability or ecological function. As a result, they tend to have low genetic diversity, making them more susceptible to pests and diseases. Because they are frequently planted out of context—poorly suited to the local climate, soil, and water conditions—they require intensive maintenance. This includes heavy applications of fertilizers, soil amendments, pesticides, herbicides, and irrigation to survive in unnatural settings.

One of the most harmful consequences of conventional planting choices is the widespread use of chemical pesticides, herbicides, and synthetic fertilizers. In addition to the effects of synthetic fertilizers on the

Plants that are bred for their ornamental value tend to need more inputs and are less disease resistant than the straight species or native plants.

eutrophication downstream water systems we discussed in chapter 4, the United States Environmental Protection Agency (EPA) estimates that approximately 78 million pounds (35,380 metric tons) of pesticides and 90 million pounds (40,823 metric tons) of herbicides are applied annually to lawns and gardens in the United States alone—an amount equivalent to the weight of 2,100 eighteen-wheeler trucks. The effects of these chemicals extend beyond the intended target species, impacting pollinators, wildlife, and even human health. According to the National Cancer Institute, children in households with pesticide-treated lawns face a 6.5 times greater risk of developing leukemia. In the European Union, pesticide use remains significant, with approximately 355,000 tons (322,000 metric tons) sold in 2022. According to Eurostat, fungicides and bactericides made up the largest share at 43 percent, followed by herbicides, haulm destructors, and moss killers at 35 percent. The highest levels of pesticide sales were recorded in France, Spain, Germany, and Italy, which together accounted for a substantial portion of total consumption.

Beyond environmental concerns, pesticide exposure has been linked to serious health risks in Europe. A report from the European Environment Agency highlights that prolonged exposure to these chemicals is associated with an increased risk of chronic illnesses, including cancer, cardiovascular diseases, respiratory conditions, and neurological disorders. These findings emphasize the need for more sustainable land management practices to reduce reliance on chemical inputs and mitigate their harmful effects. Prevent these issues by:

- **Increasing plant diversity.** Incorporate a wide range of species, especially native plants, to enhance resilience against pests and diseases while also providing essential food and habitat for pollinators and wildlife.
- **Planting with purpose.** Select species suited to the specific site conditions, considering factors like sunlight, soil type, and moisture levels to ensure long-term success.
- **Choosing multifunctional plants.** Grow species that offer food and medicinal benefits alongside their aesthetic appeal—you never know when they might come in handy!
- **Planning for climate shifts.** Introduce plants that are at the edge of their cold hardiness range to prepare for rising temperatures. In agriculture, crops traditionally grown in Mediterranean and tropical climates are already being cultivated in areas previously considered too cold.

Relying solely on ornamental plants overlooks essential human needs for food and medicine while failing to create landscapes that sustain pollinators and wildlife, including vulnerable species like bees and songbirds. A lack of plant diversity, along with dependence on excessive irrigation and synthetic inputs, weakens both our landscapes and ecosystems. As climate change intensifies droughts, pests, and extreme weather, fostering biodiversity and resilient growing practices will be key to adaptation and survival.

Anise hyssop (*Agastache foeniculum*) is a powerhouse pollinator plant beloved by honeybees. Native to North America, its long-lasting, nectar-rich blooms offer vital summer food to support healthy pollinator populations. In contrast, herbicides and pesticides degrade habitat and poison essential pollinators.

Right Maintaining a conventional lawn comes with significant environmental and financial costs. While lush, green lawns may appear pristine, they require intensive care that disrupts natural ecosystems.

THE GLOBAL COST AND IMPACT OF LAWN MAINTENANCE

Maintaining a conventional lawn comes with significant environmental and financial costs. While lush, green lawns may appear pristine and healthy, they require intensive care that disrupts natural ecosystems. Unlike the diverse plant life found in nature, lawns are monocultures—expanses of a single species—which makes them highly vulnerable to pests and diseases. Because these uniform grass stands cannot sustain themselves, they require chemical fertilizers, pesticides, and herbicides to remain viable.

The financial commitment to lawn care is immense. In the United States alone, households spend an estimated $60 billion USD annually on maintaining their lawns, including expenses for lawn care products and professional services. Meanwhile, research from NASA in 2018 estimates that American lawns span approximately 40.5 million acres (16.4 million ha), making turfgrass one of the most widely cultivated plant types in the country.

However, the environmental impact of lawn care extends beyond financial and land use concerns. A University of California, Irvine, study revealed that greenhouse gas emissions from lawn maintenance—including fertilizer and pesticide production, irrigation, mowing, and leaf blowing—are four times higher than the amount of carbon stored by grass. This means that, rather than acting as a carbon sink, lawns contribute more CO_2 to the atmosphere than they absorb.

Lawn coverage is extensive beyond the United States as well. In Canada, turfgrass occupies an estimated 7.4 million acres (3 million ha), making it one of the most widely grown plant types. Across European cities, urban green spaces, including lawns, account for 3 to 8 percent of total land area, playing a significant role in land and water use.

THE ENVIRONMENTAL TOLL OF LAWNS WORLDWIDE

A study published in *Nature Communications* found that urban lawns across Europe require up to 50 percent more water per square meter than natural grasslands, adding strain to water supplies, while research from The University of Western Australia suggests that emissions from mowing, watering, and fertilizing turfgrass may surpass its carbon sequestration capacity, echoing findings from studies out of the United States. In China, lawns and other urban green spaces account for over 40 percent of pesticide use in some metropolitan areas, raising concerns about soil and water contamination.

These findings highlight the significant global economic and environmental footprint of lawn maintenance. Sustainable landscape management, including reduced reliance on chemical inputs and alternative ground covers, could help mitigate some of these impacts.

The Plants in Your Landscape

Identifying and understanding the vegetation on your site is essential for creating a resilient landscape that supports people, wildlife, and soil health. Plants form the foundation of any ecosystem, influencing water retention, soil stability, and biodiversity. By assessing existing vegetation patterns, plant interactions, and environmental conditions, you can make informed decisions about what plants to introduce into your landscape.

WHAT SHOULD YOU LOOK FOR WHEN MAPPING EXISTING VEGETATION?

In both urban and suburban landscapes, common plant patterns and conditions reveal opportunities for improving biodiversity and long-term ecological health.

Overgrown or Aggressive Species

Some plants spread aggressively, reducing biodiversity and making landscape management more difficult. Identifying these species early helps you make informed choices about containment, removal, or replacement with more beneficial plants.

- **Sparse or struggling vegetation.** Patchy plant growth may indicate poor soil health, compaction, or lack of water infiltration.
- **Excessive lawn or ornamental plantings.** These are high-maintenance spaces with limited habitat value.
- **Dominant species.** Are you constantly weeding out a particular plant from your beds? When you look into your yard, do you see any particular plant species dominating the plant palette?

Aggressive species crowd out biodiversity in your landscape. Knowing which aggressive species you might be facing in your landscape will help you craft a maintenance strategy to either remove them or keep them at bay so other species can thrive.

Right Most landscapes contain various planting layers, just like a forest. These include the trees, shrubs, herbaceous, and ground cover layers. This landscape shows how each layer working together maximizes the growing space available.

Opposite Patterns of light in your yard will influence which plants you decide to cultivate. Accurately mapping your light availability will ensure that your plants are well suited for their locations, thereby reducing the need for intensive maintenance.

Vegetation Layers and Patterns

Mapping the layers of vegetation allows you to recognize strengths and gaps in your plant community, making it easier to plan for a balanced and functional design. Healthy landscapes feature layers of plant growth rather than a singular layer or type. Observing how plants are distributed can guide decisions about preservation, transplanting, or removal. Consider the following:

- **Tree canopy:** What trees exist on-site? How do they affect light conditions and understory plantings?
- **Shrub layer:** Are there woody plants beneath the trees? Do they provide habitat and structure for wildlife?
- **Herbaceous layer:** Which perennial or annual plants thrive here? Are they supporting pollinators and soil health?
- **Ground cover:** Is the soil covered with living plants, mulch, or exposed to erosion?

Light Availability

The tree canopy plays a critical role in defining sunlight patterns throughout the day. Marking the reach of the canopy helps you understand which areas receive full sun, partial sun, or shade. In residential settings there are other elements in your landscape—such as houses, freestanding buildings, and nearby structures or plantings—that will influence the light patterns on your site. Sun exposure affects plant choices, influencing growth rates, flowering potential, and productivity.

Light conditions can be categorized as follows:

- **Full sun:** More than five hours of direct sunlight per day
- **Partial sun:** three to five hours of direct sunlight per day
- **Shade:** Less than three hours of direct sunlight per day

Soil Moisture

Plant health is closely tied to soil moisture, and the soil moisture in any given part of your landscape will influence which plants you choose and how well they thrive. This doesn't mean that you need uniform soil moisture throughout your site, instead it means that you'll need to choose appropriate plants that respond to the existing conditions. Identifying the following zones helps guide planting choices to match site conditions.

As you think about your site, note how water moves through the landscape and how different areas retain moisture. Common conditions include:

- **Wet:** Standing water present
- **Mesic:** Well-hydrated soil with no standing water twelve hours after rainfall
- **Dry:** Well-drained soil with no standing water

Left Different plants tell a story of soil moisture in your landscape. By identifying water-loving species in your yard, such as sedges (*Carex* spp.) or ground ivy (*Glechoma hederacea*), pictured here, you'll be able to tell a lot about your soil moisture.

Opposite A plant community of loquat (*Eriobotrya japonica*) with an understory of native perennials produces food and provides seasonal interest.

Wildlife Stakeholders

What animals support and are supported by this ecosystem? This will help you identify what types of critters you'll be working with as you build your garden. Every landscape supports a network of animal life, from pollinators and birds to soil organisms and larger wildlife. Observing these interactions can help you:

- Identify species that rely on existing vegetation.
- Enhance habitat with plantings that provide food and shelter.
- Recognize potential challenges, such as overgrazing by deer or invasive pests.
- Integrate plants that support wildlife, creating a more dynamic and resilient ecosystem.

Structural Planting Zones

Many residential landscapes feature existing structural elements that influence plant distribution. Understanding how these spaces function allows you to build upon existing structures or modify them to better align with ecological goals. These include:

- Foundation plantings around buildings
- Hedgerows and privacy screens along property lines
- Border planting beds that define pathways or garden edges
- Lawn areas that may be reimagined for greater ecological function
- Woodland edges where different plant communities transition

Topographical Influence on Plant Growth

A site's position within the topography of an area affects soil composition, moisture levels, and plant adaptability. Recognizing these factors helps in selecting plants suited to the natural conditions of your landscape. Consider whether your site sits on a:

- **Ridge or slope:** Often drier, with nutrient loss due to erosion
- **Mid-slope:** A transitional area with variable moisture and soil conditions
- **Valley or lowland:** Typically richer in nutrients but prone to water retention

Pests and Plant Health

Observing plant health can provide clues about underlying site conditions. Common issues include:

- Molds, mildews, and blights, which are often linked to poor air circulation or excessive moisture.
- Pest infestations, which may indicate plant stress due to unsuitable growing conditions.
- Nutrient deficiencies, as reflected in poor growth, yellowing leaves, or weak root systems.

Supporting plant health through soil improvement, proper plant selection, and biodiversity encourages natural pest control and disease resistance. By carefully analyzing your site conditions, you can make informed decisions that lead to a more resilient, ecologically functional landscape. Thoughtful plant choices and strategic planning will help create a thriving system that benefits people, wildlife, and the land itself.

Regenerative Planting Strategies

In ecological design, plants are carefully chosen and arranged to work together, mimicking the resilience and balance of natural ecosystems, fostering biodiversity, strengthening ecological connections, and improving soil health. By selecting plants that support each other and the surrounding environment, we create self-sustaining landscapes that provide multiple benefits:

- **Restoring ecosystem functions:** Native plant communities are key to restoring degraded areas, controlling erosion, and improving soil and water quality.
- **Reducing maintenance needs:** Because these plants are well suited to local environments, they require less water, fertilizers, and pesticides compared to conventional landscaping.
- **Capturing carbon:** Plants with deep root systems help store carbon in the soil, helping to mitigate climate change by reducing CO_2 levels in the atmosphere.
- **Improving water management:** Well-planned plant communities help absorb excess rainwater, reduce runoff, and naturally filter pollutants before they reach waterways.

The reliance on ornamental plants, pesticides, and synthetic fertilizers ultimately weakens landscapes rather than strengthening them. In contrast, choosing resilient, site-appropriate, and biodiverse plant species can help restore ecosystems, improve soil quality, and minimize reliance on synthetic fertilizers and pesticides—fostering a more sustainable and regenerative approach to landscaping. In the next chapter, we will explore strategies you can use to create biodiversity in your yard while also producing food and habitat, ensuring that your garden supports both people and wildlife. We'll talk about incorporating an understanding of plant communities into ecological design, so our landscapes become more self-sufficient, resilient, and supportive of long-term environmental health. By choosing the right plants, we can create landscapes that are both beautiful and beneficial.

Queen Anne's lace (*Daucus carota*) naturalizes to add a structural layer to a client's front yard meadow. Dense, upright flowering species and grasses supply structure for more tender perennials.

CHAPTER 8

HOW TO BUILD BIODIVERSITY WHILE FEEDING YOURSELF AND WILDLIFE

A well-designed permaculture landscape does more than just produce food—it creates a thriving ecosystem that supports both people and wildlife. By choosing plants that serve multiple functions, we can build biodiversity while ensuring a continuous harvest of food, medicine, and habitat resources. Instead of traditional monoculture gardens, which deplete soil and require constant inputs, permaculture encourages diverse plant communities that work together to enrich the land, attract pollinators, and strengthen ecological resilience. In this chapter, we'll explore how to design plantings that nourish you while also supporting beneficial insects, birds, and soil life, creating a self-sustaining system that thrives year after year.

Rethinking Lawns: From Monocultures to Ecosystems

So, does this mean you have to give up having a beautiful outdoor space for your family? Not at all! The key is to shift from planting isolated species to cultivating entire ecosystems. Instead of thinking in terms of individual plants, think in terms of plant communities—diverse groups of species that support the environment in multiple ways.

WHAT IS A PLANT COMMUNITY?

Danish botanist Johannes Eugenius Bülow Warming introduced the term "plant community" in his 1895 book *Plantesamfund*, which translates to "Plant Communities" or "Plant Societies." His work laid the foundation for modern ecology by systematically studying how plant species interact with their environments.

American ecologist Frederic Clements expanded on Warming's concept of plant communities in the early twentieth century. He theorized that plant communities operate as interconnected, interdependent systems, likening them to a "superorganism." This idea highlighted the way plant communities evolve and change over time through ecological succession.

The idea of plant communities has been central to ecological studies, influencing how scientists understand species interactions, succession, and the organization of biodiversity within ecosystems. A plant

Left Echinacea (*Echinacea purpurea*), Virginia spiderwort (*Tradescantia virginiana*), and blue flag iris (*Iris versicolor*) create a beautiful and textured meadow planting.

Opposite, top A cottage garden filled with native and useful species replaces the lawn, supporting a host of pollinators, songbirds, and wildlife while growing food and medicine.

Opposite, bottom Introducing understory species, especially ones that provide berries, is a great way to support local songbirds.

community consists of a variety of species that work together to create a balanced and resilient landscape. These plants contribute to the ecosystem by distributing nutrients, offering varied habitats, supporting deeper root structures for soil health, and meeting human needs.

THE ROLE OF PLANT COMMUNITIES IN ECOLOGICAL DESIGN

Plant communities play a vital role in maintaining biodiversity by fostering a wide range of species that interact in beneficial ways. Here's how they contribute:

Supporting Wildlife

A variety of plant species creates essential resources for wildlife, supplying food, shelter, and nesting sites for insects, birds, and small mammals.

Supporting Pollinators

A variety of flowering plants ensures a consistent food source for pollinators like bees, butterflies, and hummingbirds, helping sustain their populations.

Natural Pest Management

Biodiverse plantings attract beneficial insects that keep pest populations in check, reducing the need for chemical pesticides.

Enriching Soil Health

Individual plant species contribute to soil fertility by fixing nitrogen, adding organic matter, and stabilizing the ground with varied root structures.

Permaculture and Native Plants

I frequently get asked whether permaculture promotes the exclusive use of native plants or, conversely, if it encourages the use of so-called "invasive" species. I put *invasive* in quotes because that term carries a lot of nuance and debate. This is not a simple yes-or-no question, and I'd love to dig into it with you to explore the complexities.

Understanding the distinctions between native, naturalized, and aggressive species is crucial when

selecting plants for a resilient and ecologically balanced landscape. This topic is often highly debated, with strong opinions on both ends of the spectrum. The conversation around native versus invasive plants can sometimes feel dogmatic, making it essential to explore both perspectives before making informed decisions.

I prefer to take a nuanced approach to this issue when developing planting plans. But before diving into the complexities of the debate, let's define some key terms.

Left Mountain mint (*Pycnanthemum muticum*) is a quick spreading perennial native to the mid-Atlantic region of the eastern United States. Even in its native territory, it is often mislabeled as invasive because of its ability to spread quickly. Knowing the behaviors of the plants you introduce is crucial.

Above Flowering perennials, such as this verbena (*Verbena bonariensis*) provide necessary nectar sources to pollinators.

- **Native species** are plants indigenous to a specific area, having evolved alongside local ecosystems and wildlife.
- **Naturalized species** are non-native plants that have established themselves over time, integrating into the ecosystem without significantly disrupting its balance.
- **Invasive non-native species** are those that spread aggressively, often outcompeting native plants and altering ecosystem dynamics.

The classification of a non-native plant as invasive is not always straightforward. Some ecologists argue that invasive species reduce biodiversity by crowding out native plants, disrupting natural food webs, and altering soil conditions. This perspective has shaped conservation policies at both state and federal levels, often advocating for aggressive removal efforts, including the use of chemical herbicides. The term *noxious weed*, as defined by the EPA, applies to both native and non-native species that pose a threat to monoculture crops or livestock, further complicating the conversation.

On the other hand, some ecologists challenge the traditional view of non-native invasive species, arguing that ecosystems are dynamic and continuously evolving. From this perspective, the presence of certain non-native plants can sometimes provide ecological benefits, such as soil stabilization or habitat for pollinators, particularly in degraded landscapes. David Theodoropoulos, in his book *Invasion Biology*, takes this argument further, examining the origins of invasion biology and its ties to historical and political ideologies. He controversially suggests that the war on invasive species is an extension of larger systemic forces, including industrial agriculture and militarized environmental management.

With such strong and opposing viewpoints, it's clear that the issue of invasive species is complex, requiring a thoughtful and context-based approach rather than a rigid, one-size-fits-all solution. In the following sections, we'll explore how to navigate these nuances when designing a plant palette that supports ecological resilience, biodiversity, and long-term sustainability.

Generally speaking, people are quick to anthropomorphize and demonize plants, but in my assessment, they're just doing their work. Here in Atlanta, everybody is always complaining about English ivy (*Hedera helix*), saying it kills the trees, and it chokes everything out. It's aggressive and can definitely take over if not managed. However, if you see the work it's doing within the ecosystem, it's taking these depleted, degraded, disturbed soils on the cleared edges of land and completely rebuilding topsoil. What incredible work! There is an intelligence to plants—the voids they're filling, the work they're doing, and the ways they interact—that's not captured when we're so simplistic in our terms.

Rather than using the term *invasive*, which carries militaristic connotations, I prefer *aggressive* to acknowledge that these species are responding to ecological imbalances rather than intentionally disrupting ecosystems. Nature operates with innate intelligence, and when a plant dominates an environment, it often signals a lack of natural checks and balances, such as predation or competition. The common response—eradicating these species with herbicides—often backfires, degrading soil health and creating conditions that favor the very plants being targeted. Instead of waging war on these species, we can ask deeper questions: What is this plant telling us about the land? How is it contributing to ecosystem repair? By understanding its role and working with natural processes, we can guide the landscape toward greater balance, fostering a healthier, more resilient plant community.

THE CASE FOR NATIVE PLANTS

Permaculture doesn't prescribe an exclusive focus on native plants, but one of its core principles is restoring biodiversity. Native species naturally play a crucial role in ecological resilience and offer many benefits to a landscape.

Because they are adapted to the local climate and soil conditions, native plants thrive with minimal intervention, reducing the need for irrigation, fertilizers, and chemical inputs. They have also coevolved with local insects, fungi, and bacteria, making them essential players in the food web and soil microbiome. By incorporating native plants into your landscape, you help sustain pollinators, birds, and other wildlife that depend on them.

In many urban and suburban areas, native species are often displaced by aggressive, unmanaged plants. Reintroducing them is an act of ecological restoration—helping to rebuild biodiversity and create more balanced, self-sustaining ecosystems.

THE CASE FOR NON-NATIVE PLANTS

The role of non-native plants in permaculture is a widely debated topic, with strong opinions on both sides. However, there are several key reasons why incorporating certain non-native species can be beneficial.

Globalization has made the world more interconnected, and as climate patterns shift, some native species struggle to survive in their traditional ranges. In cases where native plants are declining—such as the eastern hemlock (*Tsuga canadensis*)—some ecologists advocate for planting non-native alternatives that can fill the same ecological niche and prevent ecosystem collapse.

Many non-native plants classified as *invasive* are actually performing critical ecological functions, particularly in disturbed or degraded landscapes. These species often help stabilize soil, restore nutrients, and provide food and habitat for wildlife. Rather than viewing them as threats, some argue that we should recognize their role in nature's process of self-repair, even if it unfolds on a timeline beyond human perception.

Additionally, in a world where truly wild spaces are becoming increasingly rare, building resilient food systems means diversifying beyond traditional annual agriculture. Integrating perennial food sources—such as fruit and nut trees, berry-producing shrubs, and edible greens—creates a more stable, abundant, and climate-adapted food supply.

Ultimately, the question isn't whether we should use *only* native or *only* non-native plants, but how we can design diverse, regenerative landscapes that support both ecological and human needs. To make it simpler for you, here is the decision-making framework I use when considering introducing a plant into a landscape, assuming it also meets other criteria and serves multiple functions:

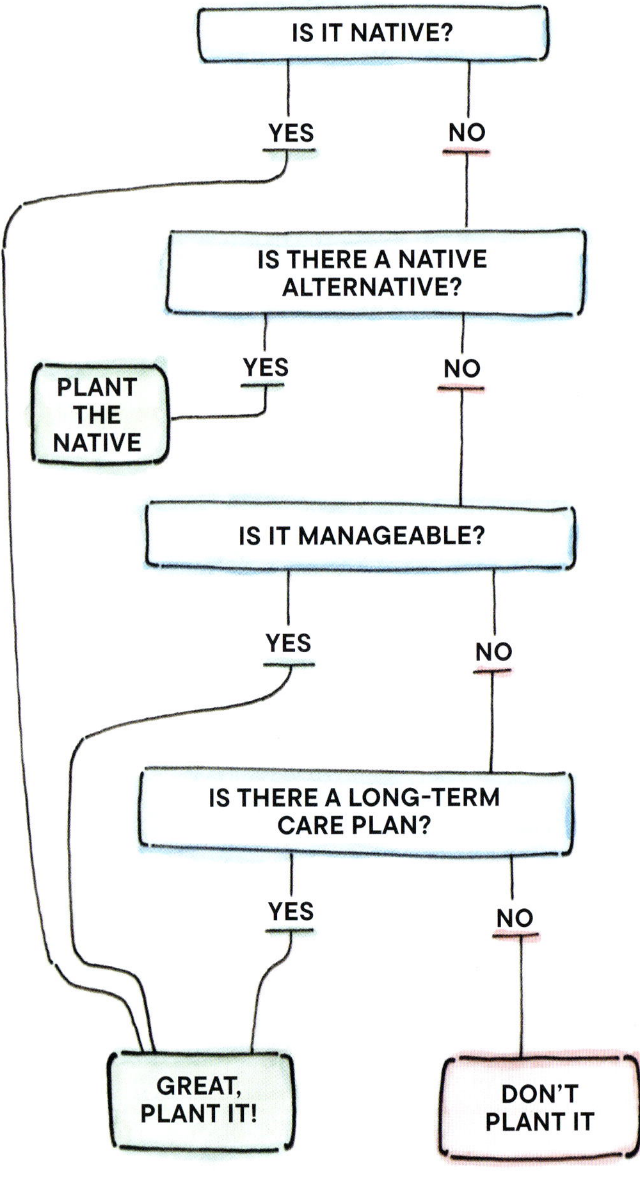

There are many nuances in determining the merits of introducing individual plants into a landscape, such as what it yields, how many functions it has within the landscape, whether or not it offers ecosystem services, which niche it fills, etc. We'll talk more about planting design in the coming chapter.

REMOVING AGGRESSIVE SPECIES

What is the best way to remove an aggressive species? This is a question that comes up quite often, and I have identified three primary methods, all of which take work and patience.

Continual Coppicing

This is where you cut the aggressive plant down to the ground, and when it sprouts again, you cut off all of the new foliage. Doing this over and over will often use up the root sugars and the plant will die back. Utilizing goats or sheep is essentially adhering to this management practice, as well, since they defoliate the plant (if it's low enough to the ground), and keep it from photosynthesizing, which kills the plant over time.

Sheet Mulching

If you do a surficial removal of the plant, you can lay a few layers of cardboard underneath mulch to keep the plant from photosynthesizing, which will eventually kill it.

Digging It Out

Generally, the most effective way to eradicate a plant is to dig it out, but it is also the most labor-intensive.

A Combination of the Above

If you're dealing with a very aggressive species, the most effective way I have found is to dig it out about four inches (10 cm) below the soil, then sheet mulch.

Opposite Since choosing a plant is not as cut and dry as saying it's native or not, here is a decision-making tree I use to determine if it's the right plant for the job.

Then, when or if it sprouts, manage it with continual coppicing (usually once per season).

A word of caution: Nature abhors a vacuum. Don't begin removing aggressive species until you have a plan in place to replace or continue managing it, so you don't create more work for yourself than necessary.

Planting for Ecological Succession

When it comes to planning your permaculture landscape, many people put "low-maintenance" on their wish list. If that's a priority for you, it is important to understand succession and disturbance dynamics.

WHAT IS ECOLOGICAL SUCCESSION?

Ecological succession is the natural process by which plant and animal communities change and develop over time. It begins when an area is disturbed—whether by fire, flood, erosion, or human activity like mowing or tilling—and then gradually moves toward a more stable and diverse ecosystem. As various species establish, compete, and interact, the landscape shifts, with fast-growing pioneer plants giving way to longer-lived species. Eventually, the system reaches a balanced state, known as a climax ecosystem, where plants and organisms exist in relative harmony with their environment. The direction of succession depends on regional climate and soil conditions—for example, an abandoned field may become a forest, while a coastal dune may stabilize into grassland. Understanding succession helps gardeners and land stewards work with nature rather than constantly battling against its inevitable progression.

The trajectory of ecological succession begins with disturbance and ends with stabilization. Disturbance includes fire, flood, landslide, erosion, clearing, tilling, mowing and any other event that sets the vegetative growth back to ground level, so to speak.

Ecological gardening is an oscillation between succession and disturbance. This is relevant to your gardening decisions, because as you plan for your long-term landscape, it is helpful to know where nature

PLANT TERMS TO KNOW

Understanding basic plant classifications and certain ecological terms helps in making informed choices. Here are three key terms to know:

Annual

An annual plant completes its life cycle within a single year, meaning it grows, flowers, sets seed, and dies. Some annuals, known as self-seeding annuals, drop seeds that sprout the following season, allowing them to reappear naturally. Example: Tomatoes (*Solanum lycopersicum*).

Perennial

Unlike annuals, perennials return year after year. While they die back to the ground in winter, their roots remain alive, producing new growth each spring. Examples: echinacea (*Echinacea* spp.) and black-eyed Susans (*Rudbeckia hirta*).

Niche

In ecology, a niche refers to the specific role a plant or organism plays within its environment, including its preferred conditions and interactions with other species. When designing plant communities, understanding niches helps create balanced, self-sustaining ecosystems.

A field in mid-succession.

is headed if left to her own devices. In order to use succession as an asset in your garden, you should look at wild spaces in your area to determine the climax ecosystem. In the temperate zones of the southeastern United States, it is usually forest. In the midwestern United States, it is grassland, and so on. Nature's momentum is moving toward that expression. For example, if you brush hog a field or stop mowing your lawn, what happens? Here in the southeast, you begin to see early succession species move in, such as sweetgum and poplar saplings. The field or lawn is moving toward reforestation.

HOW DOES SUCCESSION IMPACT MAINTENANCE?

If you are trying to maintain your garden in an entirely "disturbed" state, meaning thwarting the natural trajectory toward ecological stabilization or climax ecosystem, you are creating a lot of work for yourself. Disturbance, in most residential landscapes, is maintained through either mowing or weeding. On the other hand, without disturbance, you can see a flourishing of aggressive species or feel like your house is about to be consumed by plants.

FOREST GARDENING: UTILIZING BOTH SUCCESSION AND DISTURBANCE IN PLANNING YOUR LANDSCAPE

Designing your ecological, edible landscape with succession in mind is often called "forest gardening." Forest gardening is a way of designing food-producing landscapes that mimic natural ecosystems. By incorporating the seven layers of a forest (see sidebar page 137), this approach creates a diverse, self-sustaining system that produces fruits, nuts, berries, perennial vegetables, and medicinal and culinary herbs—while also reducing maintenance and increasing resilience. But what makes forest gardening particularly effective is that it harnesses the natural process of ecological succession rather than fighting against it.

Forest gardening is utilizing the layers of the forest to grow fruits, nuts, berries, perennial vegetables, and culinary and medicinal herbs. Incorporating the seven layers of a forest into your planting plan reduces

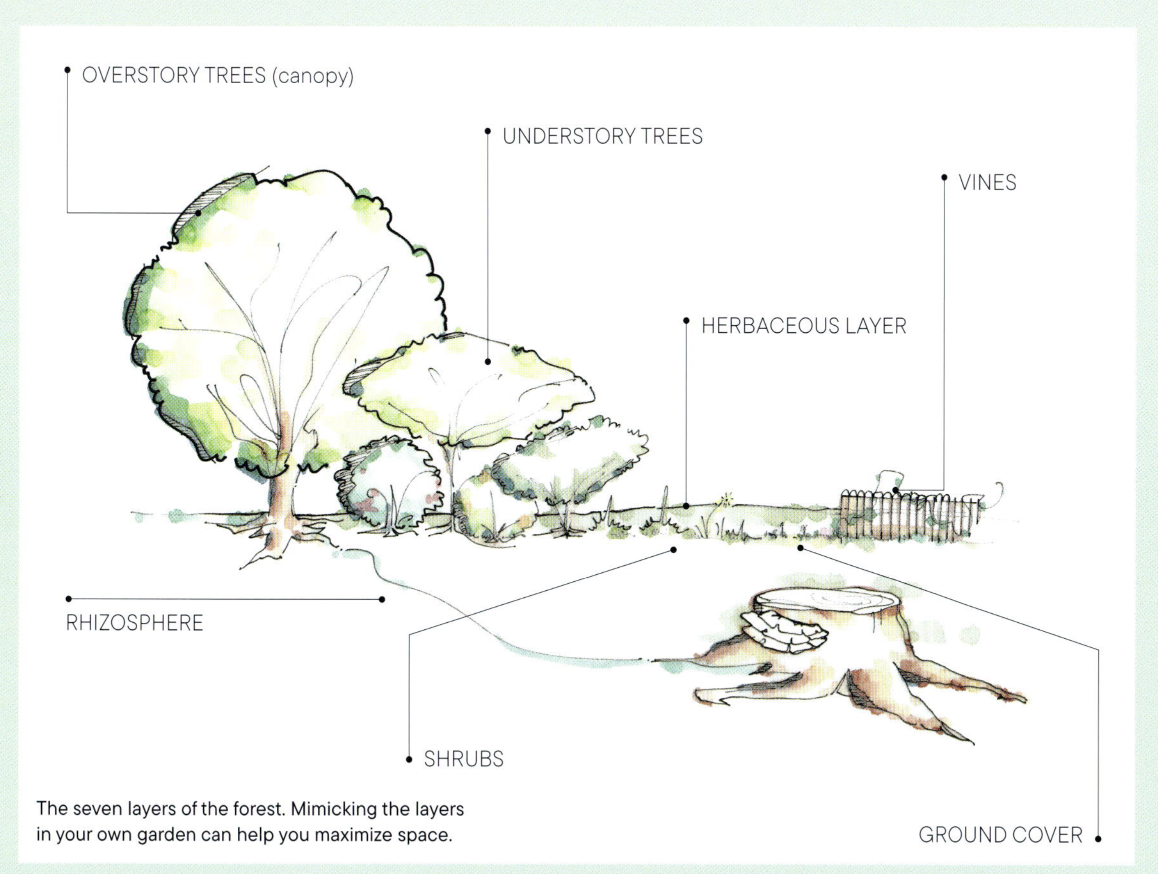

The seven layers of the forest. Mimicking the layers in your own garden can help you maximize space.

LAYERS OF THE FOREST

Understanding the natural layers of a temperate or deciduous forest can help you assess your site's vegetation and guide your plant choices. Each layer offers unique growing opportunities and ecological benefits:

- **Overstory trees (canopy):** Typically 40 feet+ (12 m+), forming the highest layer of the forest.
- **Understory trees:** Ranging from 15 to 40 feet (5 to 12 m), these trees thrive in partial shade beneath the canopy.
- **Shrubs:** Typically 6 to 20 feet (2 to 6 m), providing structure, habitat, and food sources.
- **Herbaceous layer:** Includes ferns, grasses, and forbs, growing between 1 to 6 feet (30 to 183 cm).
- **Vines:** Climbing plants that weave through trees and shrubs.
- **Ground cover:** Low-growing plants under 1 foot (30 cm), helping to retain moisture and prevent erosion.
- **Rhizosphere:** The unseen underground layer where roots, fungi, and microbes interact to support plant health.

By incorporating plants at different levels, you can create a diverse, productive, and ecologically balanced landscape.

Growing fruit trees, such as this loquat (*Eriobotrya japonica*), with useful understory species makes for a beautiful food forest in this residential garden instead of empty lawn space.

maintenance, increases productivity, and utilizes nature's momentum. Succession is nature's way of healing and stabilizing landscapes over time.

After a disturbance—such as a mown lawn, a cleared field, a burned forest, or abandoned farmland—plants naturally begin recolonizing the space in stages. The more disturbed areas you try to maintain at a static level in your landscape, the more maintenance you will have to perform over time to keep this level of disturbance. Imagine a lawn. When you stop mowing, it begins to colonize with other plants, and moves toward a new ecosystem, such as tall grassland or forest. Fast-growing pioneer species like grasses and shrubs appear first, followed by small trees and, eventually, a mature, stable ecosystem such as a forest. Each stage builds upon the last, improving soil structure, increasing biodiversity, and creating the conditions for more complex plant communities to thrive. But in order to keep the landscape at a disturbance level, you'll have to continue to mow it.

Traditional gardening and farming often work against this momentum, requiring constant effort to keep nature in a perpetual state of early succession. Lawns need mowing, annual gardens require replanting, and weeding is a never-ending battle. Forest gardening takes the opposite approach—it works with nature's trajectory instead of resisting it.

How Forest Gardening Reduces Maintenance and Increases Productivity

By designing plantings that align with natural succession, a forest garden minimizes human intervention while maximizing ecological function. Here's how:

Soil Improvement Happens Naturally

In a forest, plants drop leaves, stems, and organic matter that decompose and enrich the soil. A forest garden mimics this, building fertility over time without tilling or synthetic fertilizers. Deep-rooted plants, such as comfrey (*Symphytum officinale*), act as dynamic accumulators, pulling nutrients from deep within the soil and making them available to other plants.

Pest and Weed Management Is Built-In

In a diverse system, beneficial insects, birds, and soil microbes help naturally control pests, reducing the need for chemical intervention. A multilayered planting strategy shades out weeds, making it harder for unwanted plants to establish.

Water Retention and Climate Resilience

Tree canopies and perennial plant roots reduce evaporation and improve moisture retention, meaning less irrigation is needed. The varied heights of plant layers create a microclimate, buffering extreme temperatures and wind.

Perennials Do the Work for You

Unlike annual vegetable gardens that need replanting each year, perennial crops return season after season, reducing labor while maintaining productivity. Trees like chestnut (*Castanea* spp.) and hazelnut (*Corylus* spp.) provide nut crops with minimal upkeep. Berry-producing shrubs like currants (*Ribes* spp.) and elderberries (*Sambucus* spp.) offer food with little ongoing maintenance.

Biodiversity Supports Long-Term Stability

A forest garden attracts pollinators, beneficial insects, and wildlife, enhancing ecosystem balance. Diversity ensures that if one species struggles due to pests or climate fluctuations, others will thrive, providing built-in resilience.

Rather than forcing landscapes into a state of constant disruption, forest gardening embraces nature's trajectory toward stability. By planting in layers and selecting species that work together, we create low-maintenance, highly productive landscapes that require less watering, weeding, and fertilizing over time. Instead of battling succession, we guide it—allowing plants to fill their ecological roles while providing an abundance of food, medicine, and habitat. Forest gardening is not just a method; it's a shift in perspective. By aligning with natural processes, we transform our gardens into self-sustaining ecosystems, reducing work while increasing yields and biodiversity. Whether on a small backyard scale or a larger homestead, this approach makes for a more resilient, regenerative, and harmonious way to grow food and steward the land.

Planting Strategies for Natural Pest Management

Plant diversity is one of the best defenses against pests and disease. By avoiding monoculture plantings and incorporating a variety of species, you create a more resilient ecosystem with natural checks and balances. However, as climate patterns shift and winters become milder, pests and plant diseases are overwintering easier and persisting for longer periods. Warmer temperatures and increased spring moisture create ideal conditions for infestations and fungal growth.

To manage pests naturally, focus on prevention and ecologically mindful design rather than chemical treatments.

PREVENTIVE STRATEGIES FOR PEST AND DISEASE CONTROL

Observe Your Plants Regularly

Check leaves for insect eggs, damage, or signs of disease. Early detection makes management easier.

Target Moisture-Retaining Areas

Shady spots and north-facing spaces are often the first places where pests and fungi appear. Proper spacing and pruning reduce humidity and help prevent fungal growth.

Use Natural Sprays When Needed

A simple neem oil solution can help manage early infestations:

- 1 tablespoon (15 ml) cold-pressed neem oil
- 1 teaspoon (5 ml) liquid castile soap
- 1 gallon (3.8 L) water

Apply weekly, or twice a week for severe infestations.

Hand-Wash Affected Plants

Gently wash leaves to remove the fungus and improve plant health.

BUILDING BIODIVERSITY FOR LONG-TERM PEST MANAGEMENT

The best way to control pests naturally is by supporting a diverse ecosystem that keeps populations in check. Consider these strategies:

Attract Beneficial Insects

Ladybugs, parasitic wasps, and birds naturally manage pest populations.

Plant a Variety of Species

Monocultures invite pests; diverse plantings create a balanced environment.

Maintain Healthy Soil

Strong plants are less vulnerable to disease and infestations.

USING YOUR PLANTS FOR WINTER WILDLIFE SUPPORT

Harvesting vines to make a wreath, then weaving in trimmings from the garden, makes a beautiful way to attract birds to your garden while using material you might otherwise discard at the end of the season.

PROJECT 1: FEEDER WREATH

Create a beautiful, functional bird feeder using natural materials while also providing a visually interesting winter display for your garden.

Materials Needed

- → Grapevine or another thick, pliable vine
- → Twine or rope (cut into 3 pieces, 4 feet [1.2 m] each)
- → Seed heads, dried flowers, pinecones, and grasses

Instructions

1. **Form the wreath.** Shape the vine into a circular wreath and secure it by weaving the ends together.

2. **Prepare for hanging.** Lay the wreath flat and tie one piece of twine around it at three evenly spaced intervals.

3. **Create a hanger.** Gather the loose ends of the twine above the wreath and tie them together to form a loop for hanging.

4. **Decorate and adorn.** Attach dried flowers, seed heads, pinecones, and grasses to provide food and perches for birds.

5. **Hang and enjoy.** Display your feeder wreath in a tree or garden space where birds can easily access it. Refresh the decorations seasonally.

Gathering bundles of dried flowers in the winter is a great way to feed birds and provide habitat for wildlife while celebrating nature's beauty!

PROJECT 2: OUTDOOR DRIED FLOWER ARRANGEMENT

Materials Needed

- → Dried flowers, seed heads, and grasses
- → Twine or natural string

Instructions

1. **Collect and bundle.** Gather dried flower stems, seed heads, and grasses from your garden.

2. **Tie and secure.** Bind them together with twine to form an arrangement.

3. **Choose a display spot.** Hang or place the bundle in an outdoor space where it adds visual interest while remaining accessible to wildlife.

4. **Let nature take over.** Over time, birds and other animals will forage from the arrangement, naturally dispersing the seeds and repurposing materials.

Both projects provide food and shelter for wildlife while celebrating the beauty of nature throughout the seasons!

Left Indian blanket (*Gaillardia pulchella*) thrives with minimal water and adds a pop of bright red.

Opposite, top Espaliered fruit tree along a client's fence.

Opposite, bottom Lavender (*Lavandula* spp.) helps keep mosquitoes away with their aromatic volatile oils.

WHAT TYPES OF PLANTS TO USE

Native Plants

- Trees, shrubs, perennials, vines, and ground covers adapted to your local soil and climate.
- Are essential for supporting pollinators, beneficial insects, and wildlife.
- Require less irrigation, fertilizers, and pesticides.
- Examples: Oak (*Quercus* spp.), serviceberry (*Amelanchier* spp.), milkweed (*Asclepias* spp.), goldenrod (*Solidago* spp.).

Edible Plants

- Fruit trees (e.g., apple (*Malus domestica*), pear (*Pyrus* spp.), citrus (*Citrus* spp.), persimmon (*Diospyros* spp.))
- Nut trees (e.g., chestnut (*Castanea* spp.), hazelnut (*Corylus* spp.), walnut (*Juglans* spp.))
- Berry bushes (e.g., blueberry (*Vaccinium* spp.), currants (*Ribes* spp.), raspberries (*Rubus* spp.), elderberries (*Sambucus* spp.))
- Perennial vegetables (e.g., asparagus (*Asparagus officinalis*), rhubarb (*Rheum rhabarbarum*), artichokes (*Cynara cardunculus* var. *scolymus*))
- Annual vegetables (e.g., tomatoes (*Solanum lycopersicum*), peppers (*Capsicum* spp.), squash (*Cucurbita* spp.))
- Culinary & medicinal herbs (e.g., oregano (*Origanum vulgare*), rosemary (*Salvia rosmarinus*), chamomile (*Matricaria chamomilla*), yarrow (*Achillea millefolium*))

Useful Plants

- Cut flowers (for aesthetics, pollinators, and bouquets)
- Fiber and craft materials (e.g., hemp (*Cannabis sativa*), flax (*Linum usitatissimum*), dye plants)
- Fuel and firewood species (e.g., fast-growing coppicing trees like willow (*Salix* spp.) and black locust (*Robinia pseudoacacia*))
- Living fences and windbreaks (e.g., hedgerows, bamboo (*Phyllostachys* spp.), and willow (*Salix* spp.))

By selecting plants that serve multiple roles and placing them in intentional combinations, you can create a productive, biodiverse landscape that works with nature instead of against it. A well-designed polyculture not only provides food but also enriches the soil, reduces maintenance, and fosters a thriving ecosystem—all while making your garden more beautiful and resilient. In the next chapter, we'll discuss specific techniques you can employ in your landscape to grow useful and native plants and create planting areas that serve multiple functions for you and your family, including strategies such as polyculture lawns, vegetable gardening methods, and winter garden maintenance. These techniques will help you establish a resilient, self-sustaining plant ecosystem that thrives throughout the year.

Pomegranate (*Punica granatum*) thrives in most temperate climates without much irrigation, while yielding an abundance of delicious fruit.

CHAPTER 9

PLANT TECHNIQUES

Understanding why we need biodiverse, regenerative plant communities is one thing—knowing how to create them is another. This chapter is where vision meets action. Now that we've explored the importance of diverse plant ecosystems, it's time to dive into the practical techniques that will help you design a resilient, productive landscape that thrives with minimal intervention.

In this chapter, we'll cover key planting strategies that mimic natural ecosystems while providing food, medicine, and habitat. You'll learn how to replace a high-maintenance lawn with a polyculture of useful, low-input plants, experiment with vegetable gardening methods that regenerate soil rather than deplete it, and discover seasonal maintenance techniques to keep your garden thriving year-round.

No matter the size of your space or your level of experience, these techniques will help you create a landscape that works for you, rather than one you constantly have to work for. By harnessing the power of nature, you'll be planting not just for today, but for a healthier, more abundant future.

Creating a Polyculture Lawn

Lawns have long been a defining feature of residential landscapes but maintaining that perfectly manicured green carpet comes at a steep price—both financially and environmentally. Between sod, seed, fertilizers, pesticides, mowing, and leaf blowing, homeowners collectively spend thousands of dollars each year on lawn care. Yet from an ecological perspective, conventional lawns are among the least sustainable landscaping choices.

So, what's the alternative? You could replace your lawn with a food garden, a wildflower meadow, or even an orchard—all fantastic options! However, the reality is that open green spaces serve a functional purpose, whether for recreation, pets, or aesthetic balance. Instead of eliminating lawns altogether, why not reimagine them?

While lawns themselves are not inherently "bad," lawn care as we know it is due for a transformation. As we've discussed, the conventional approach to landscaping often prioritizes aesthetics over ecology, leading to the widespread use of non-native species,

chemical inputs, and practices that degrade soil health and biodiversity. By rethinking our relationship with lawns, we can create landscapes that are healthier, more resilient, and better for the environment—all while maintaining beauty and functionality.

A polyculture, or lazy lawn, is an eco-friendly, low-maintenance alternative to a traditional turf lawn. Instead of a single species of grass that requires constant mowing, watering, and chemical treatments, a lazy lawn incorporates a diverse mix of low-growing, self-sustaining plants that thrive with minimal intervention.

WHY RETHINK LAWN CARE?

A more ecological approach to lawn care benefits not only your landscape but also the greater environment. Here's why it matters:

1. Restoring habitat with native plants
2. Encouraging biodiversity
3. Improving water management
4. Reducing chemical inputs

Each of these points is interconnected, but let's explore them individually to understand their importance.

OBSERVE AND INTERACT: LEARN FROM THE EXISTING PLANTS ON YOUR SITE

Before planting anything new, take a close look at what's already growing in your yard. Existing plants provide clues about soil health, moisture levels, and what species are naturally suited to your site.

Step 1: Identify Key Trees and Plants

- Pine (*Pinus* spp.) and red oak (*Quercus rubra*) create acidic soil, which is great for blueberries and azaleas.
- Beech (*Fagus grandifolia*) and tulip poplar (*Liriodendron tulipifera*) grow in fertile, well-draining soil, indicating good organic matter.
- Aggressive species like privet (*Ligustrum sinense*) or English ivy (*Hedera helix*) often thrive in disturbed areas and outcompete native plants. If you plan to remove them, have a plan for replanting immediately.

Step 2: Check for Plant Health

- Thriving plants: These indicate conditions that are ideal for similar species.
- Struggling plants: Wilting, yellowing, or stunted growth can signal poor soil, too much or too little water, or nutrient deficiencies.
- Bare patches: These may indicate soil compaction, erosion, or areas that receive too little light for most plants.

Step 3: Use Observations to Plan Your Garden

- Replace invasive non-native plants with beneficial species: Choose native plants that support pollinators and improve soil health.
- Mimic what's already working: If a certain species is thriving, plant more of it or similar plants nearby.
- Match plants to their ideal conditions: Place moisture-loving plants in damp areas and drought-tolerant species in drier spots.

By learning from what's already growing, you can design a garden that requires less maintenance while supporting a healthier ecosystem.

Opposite A polyculture lawn incorporates a diversity of species that maximize the light and moisture differences in a yard, while providing flowers for pollinators and reducing the need for mowing and irrigation.

Right Maypop (*Passiflora incarnata*), native to the southeastern United States, supports biodiversity by attracting pollinators like bees and butterflies, serving as a host plant for fritillary caterpillars, providing fruit for birds and small mammals, and stabilizing soil with its sprawling vines.

Restoring Habitat with Native Plants

When people hear about the decline of pollinators, the conversation often focuses on saving the bees. But the real crisis goes beyond honeybees, which are native to Europe. Many native pollinators, particularly here in North America, including specialized bees and butterflies, are disappearing due to habitat loss. Unlike honeybees, which can feed on a variety of flowers, native pollinators often rely on specific host plants to survive.

Why Native Plants Matter

- **Better for pollinators.** Native plants produce nectar and pollen that are more nutritious for local insects.
- **Essential for caterpillars.** Many butterfly and moth species require specific native plants for their larvae to feed on.
- **Prevents food deserts for wildlife.** Replacing native ground covers with non-native turf grasses creates a landscape devoid of essential food sources.

For example, the spangled fritillary butterfly (*Speyeria cybele*) relies on violets (*Viola sororia*) as a host plant. When landscapes are dominated by Bermuda grass or zoysia, these caterpillars have no food source, leading to population declines. This concept, known as trophic cascade, impacts entire food webs—fewer caterpillars mean fewer birds and fewer predators, disrupting the ecological balance.

Encouraging Biodiversity

Traditional lawns prioritize uniformity, often consisting of a single species of grass maintained through mowing and chemical treatments. This lack of diversity creates fragile ecosystems that are highly dependent on human intervention.

Why Biodiversity Is Key

- **Increases resistance to pests and disease.** A diverse plant mix ensures that if one species is affected by a pest or disease, others can thrive.
- **Healthier soils.** Different plant species release unique root exudates, feeding diverse soil microbes and improving soil structure.
- **Creates a self-sustaining system.** The more plant species present, the more complex and resilient the ecosystem becomes.

How Plants Build Healthy Soil

Plants interact with soil life in a mutually beneficial relationship:

- Through photosynthesis, they produce sugars.
- Some of these sugars are released through their roots, feeding bacteria, fungi, and other microbes in the soil.
- In return, soil microbes break down organic matter, making nutrients available to plants.

By increasing the variety of plant species in your lawn, you foster a thriving underground ecosystem that improves water infiltration, soil aeration, and nutrient availability—all without synthetic fertilizers.

Improving Water Management

Healthy soils are porous, allowing water to infiltrate rather than run off. However, compacted lawns with shallow-rooted grasses contribute to poor water absorption, leading to flooding, erosion, and runoff pollution.

How Plant Choices Affect Water Infiltration

- Deep-rooted species like yarrow (*Achillea millefolium*) can grow 1 to 2 feet (30 to 61 cm) deep, improving soil aeration and drought resistance.
- Turf grasses like Bermuda grass (*Cynodon dactylon*) have shallow, fibrous roots, limiting water infiltration and requiring frequent irrigation.
- Diverse ground cover plants create a network of root structures that increase soil porosity and reduce runoff.

Sod Farming and Runoff

Many conventional lawns are established using sod, which may seem convenient but has unintended consequences:

- Sod farms strip large areas of soil, leaving behind bare, compacted land that struggles to absorb water.
- When installed, sod often sits on compacted ground, reducing its ability to absorb rainfall.
- This results in more stormwater runoff, carrying fertilizers and pesticides into waterways.

By incorporating deep-rooted native grasses and perennial ground covers, you can improve water infiltration, reduce flooding, and create a landscape that thrives with minimal irrigation.

Reducing Chemical Inputs

Conventional lawn care often relies on synthetic fertilizers, herbicides, and pesticides, which have harmful environmental and health effects.

Opposite, left In North America, monarch butterflies, seen here on *Allium* spp., depend on native milkweeds (*Asclepias* spp.) for reproduction and nectar-rich wildflowers for energy during migration, but habitat loss and pesticides have led to drastic population declines. Restoring native plants supports monarchs, strengthens ecosystems, and helps preserve this iconic species.

Opposite, middle Butterfly milkweed (*Asclepias tuberosa*) adds a gorgeous pop of orange to a pollinator garden while providing an essential nectar source for monarch butterflies.

Opposite, right Diverse ground cover plants create a network of root structures that increase soil porosity and reduce runoff.

The Problem with Lawn Chemicals

- Nitrogen fertilizers contribute to algae blooms in waterways, damaging aquatic ecosystems.
- Pre-emergent herbicides prevent weed germination but also harm beneficial soil microbes.
- Pesticides used on lawns can be toxic to birds, pollinators, and even pets.

The Carbon Footprint of Lawn Care

Chemical fertilizers and herbicides are petroleum-based products with high carbon footprints. Reducing their use is one of the simplest ways to lower your landscape's environmental impact.

Health Risks to Humans and Pets

Some lawn chemicals have been linked to cancer, skin conditions, and respiratory issues. Many veterinarians have noted correlations between chemically treated lawns and pet illnesses, including skin cancer. By shifting to organic lawn care practices, you can reduce exposure to harmful chemicals while fostering a naturally resilient landscape.

DESIGNING A POLYCULTURE LAWN

The key to a successful polyculture lawn is choosing plant species that complement each other and serve multiple functions. Species selection will depend on where you live, the sun/shade conditions, soil type, and the intended use of the space. Instead of relying on high-maintenance turf grass, consider low-growing native ground covers and resilient perennials.

Recommended Ground Covers and Low-Maintenance Lawn Alternatives

- St. Augustine grass (*Stenotaphrum secundatum*): A tough, native grass that handles foot traffic.
- Common violet (*Viola sororia*): A keystone species providing food for caterpillars and early-season pollinators.
- Yarrow (*Achillea millefolium*): A deep-rooted, drought-tolerant perennial that withstands mowing.
- Blue-eyed grass (*Sisyrinchium angustifolium*): A resilient, low-growing native with charming blue flowers.
- Pony's foot (*Dichondra carolinensis*): A spreading, evergreen ground cover that thrives in tough conditions.

These ground covers are best suited for mesic to slightly xeric conditions—meaning they thrive in soils that are moderately moist but can tolerate some drought once established. They are ideal for regions with mild to warm summers and moderate winters.

- Ideal Climate Range:
  - Winter Minimum: 10°F / –12°C (USDA Zone 8)
  - Summer Maximum: 100°F / 38°C
- Best Regions:
  - Southeastern and South-Central United States (e.g., Georgia, Texas, Florida, South Carolina)
  - Coastal and temperate parts of Australia (e.g., New South Wales, southern Queensland, parts of South Australia)
  - Milder areas of the United Kingdom with good summer sunlight and low foot traffic

Perennial red fescue (*Festuca rubra*) is native to temperate regions of the Northern Hemisphere, including North America, Europe, and Asia. It thrives in cool, moist climates, often found in meadows, coastal dunes, woodland edges, and alpine regions. It makes a great base for overseeding other useful species and can handle dappled light with minimal irrigation or mowing.

Perennial Lawn Combinations:

- Perennial fescue + yarrow (*Achillea millefolium*) + dandelion (*Taraxacum officinale*) + clover (*Trifolium repens*) = A classic mix that improves soil health and resists pests.
- Clover (*Trifolium repens*) + creeping jenny (*Lysimachia nummularia*) + low-growing St. John's wort (*Hypericum perforatum*) = A colorful, resilient combination that attracts pollinators.
- Buffalo grass (*Bouteloua dactyloides*) + wild strawberry (*Fragaria virginiana*) + self-heal (*Prunella vulgaris*) = A mix that provides food for both people and wildlife.
- Blue-eyed grass (*Sisyrinchium angustifolium*) + common violet (*Viola sororia*) + native sedges (*Carex* spp.) = A shade-tolerant alternative for tree-covered areas.

These combinations are adaptable and thrive in mesic to moderately xeric conditions. They favor cool to temperate climates and include species that are widely naturalized or native across North America, the United Kingdom, and temperate regions of Australia. Most are perennial and can handle a range of soil types with minimal inputs.

- Ideal Climate Range:
  - Winter Minimum: –20°F / –29°C (USDA Zone 5)
  - Summer Maximum: 95°F / 35°C
- Best Regions:
  - Much of the continental United States, especially the Northeast, Midwest, and Pacific Northwest
  - Cooler parts of Australia (e.g., Victoria, Tasmania, elevated regions of New South Wales)
  - Most of the United Kingdom, especially where lawns receive some sun and moisture

Note: *While these combinations are generally hardy, success depends on matching the right species mix to your site's specific light, moisture, and foot-traffic conditions. For instance, buffalo grass prefers full sun and well-drained soils while blue-eyed grass and violets do well in partial shade.*

Rethinking our approach to lawn care doesn't mean sacrificing beauty or function—it means working with nature instead of against it. By incorporating native plants, increasing biodiversity, improving soil health, and reducing chemical inputs, you create a low-maintenance, ecologically beneficial landscape that is beautiful, functional, and environmentally responsible.

A polyculture lawn isn't just better for the planet—it's also easier on your time and budget. By moving away from high-maintenance monoculture lawns and embracing low-growing, diverse plantings, you can still enjoy open space. After all, who wouldn't want a lawn that looks great, supports the environment, and requires less work and money to maintain? Polyculture lawns are the perfect solution for anyone looking to balance beauty, function, and sustainability in their yard.

Veggie Gardening Methods

A big component of a permaculture landscape is that it grows food for humans and wildlife. While we have spent a lot of time focusing on perennial plantings and methods, we all know that annual vegetables are a large part of the modern human diet. So, naturally, growing a vegetable garden is an important element in any permaculture landscape.

Successfully growing a vegetable garden starts with selecting the right type of garden bed. Each method has its benefits and challenges, so it's important to consider your space, budget, and long-term goals before making a decision. Whether you're looking for a no-cost, eco-friendly approach or a structured raised bed system, there's a solution to fit every garden. Below, I explore different garden bed options, highlighting their pros and cons to help you decide which method (or combination of methods) works best for you.

HÜGELKULTUR BEDS

For a cost-effective, water-retaining, and soil-building vegetable garden bed, *hügelkultur* is a fantastic choice. As previously discussed in chapter 6, on page 106, *hügelkultur* beds are raised beds built with rotting logs, branches, and organic material, which decompose over time to improve soil fertility and retain moisture. The wood acts like a sponge, soaking up rainwater and slowly releasing it to plant roots. This method drastically reduces irrigation needs and can be customized in height for accessibility and vertical gardening. *Hügelkultur* beds are particularly useful for sloped areas and erosion control.

However, they require some labor upfront, and sourcing logs may take effort. They're best suited for herbaceous plants like vegetables, flowers, and herbs, rather than trees and shrubs, as the bed settles over time.

Pros

- Reduces the need for irrigation
- Improves soil fertility over time
- Helps control erosion
- Can be built affordably with natural materials

Cons

- Labor-intensive to construct
- Requires access to logs or wood debris

RAISED BEDS

For a structured, easy-to-maintain growing space, raised beds are a classic choice. Raised beds are aesthetic, convenient, and customizable. They provide better drainage, keep grass and rhizomatous weeds out, and allow for higher-quality soil. These beds are commonly built from wood, stone, or metal, and can be raised to different heights for easier access and less bending while gardening.

However, they can be costly to build, especially if using high-quality materials. Additionally, raised beds require an initial soil investment and may need regular irrigation due to their elevated exposure to heat and air. To improve moisture retention and nutrient cycling, consider placing a layer of decomposing logs (a mini *hügelkultur* bed) at the bottom or integrating organic matter like compost and mulch.

Pros

- Easy to maintain and weed
- Keeps grass and rhizomatous weeds out
- Customizable height for convenience

Cons

- Can be expensive to build
- Requires initial soil investment
- May need irrigation and soil amendments
- Choosing the right raised bed materials

IN-GROUND BEDS

For an easy and affordable way to start gardening, in-ground beds are a great option. In-ground beds work best in flat or gently sloped areas. They require minimal effort to establish—simply remove grass and weeds, amend the soil, and start planting.

There are a few methods for preparing an in-ground garden:

- Solarization: Covering the soil with clear plastic to trap heat and kill grass and weeds.
- Occultation: Using a black tarp to block light and smother weeds.
- Sheet mulching: Layering cardboard, compost, and mulch to naturally suppress grass while enriching the soil

While low-cost and water-efficient, in-ground beds may require more frequent weeding and soil amendments compared to raised beds.

Pros

- Simple and inexpensive to start.
- Uses existing soil, reducing input costs.
- Less irrigation required than raised beds.

Cons

- Can be more labor-intensive to maintain.
- More susceptible to weeds.

MATERIALS FOR RAISED BEDS

When building raised beds, the material you choose impacts cost, longevity, and sustainability.

Wood

- Best choice: Naturally rot-resistant species like black locust (*Robinia pseudoacacia*) or cedar (*Thuja plicata*).
- Most accessible: Pressure-treated pine, though some gardeners avoid it due to past concerns about chemicals (modern versions are CCA-free).
- Downsides: Wood absorbs moisture, potentially increasing irrigation needs.

Stone, Brick, and Masonry

- Long-lasting and visually appealing.
- Retains heat, which may overheat soil in warmer climates.
- More expensive and can be labor-intensive to install.

Metal

- Durable and modern-looking, but it can increase soil temperature quickly.
- Galvanized steel is a popular option but should be lined if using for food gardens to prevent leaching.

Opposite A mounded *hügelkultur* bed grows annual vegetables without the need for extensive irrigation.

Above A terraced raised bed creates a weed barrier with the edge and allows the soil to be built up. The height of raised beds can vary depending on your aesthetic and mobility preferences.

Opposite In-ground beds allow you to plant directly in the soil without the added costs of additional planting medium or construction materials.

Right Maximize space by incorporating vegetables and edible flowers into existing ornamental beds. If you have limited sun or space, pocket gardening is an easy way to integrate food production into your existing landscape.

POCKET GARDENS IN EXISTING PLANTING BEDS

Maximize space by incorporating vegetables and edible flowers into existing ornamental beds. If you have limited sun or space, pocket gardening is an easy way to integrate food production into your existing landscape. Annual vegetables, herbs, and edible flowers can be tucked between perennials and shrubs, creating a beautiful, functional garden. However, this method requires careful plant placement and a keen eye for identifying different species, since edibles will be intermingled with ornamentals. Additionally, scattered plantings may be harder to protect from pests and wildlife.

Pros

- Makes use of existing planting space.
- Adds seasonal color and diversity.
- Can reduce irrigation needs.

Cons

- More difficult to manage and harvest.
- Requires more plant identification skills.
- Can be vulnerable to pests.

TERRACE BEDS FOR SLOPED LAND

For gardeners working with sloped terrain, terrace beds create level growing spaces while preventing erosion. Terraces stabilize slopes, prevent water runoff, and create usable planting areas on otherwise difficult land. Instead of building a full box, most terraces require only three sides, reducing material costs. However, terraces can be tricky to access if built too high or too steep, so they should be designed with walkways or steps for easy mobility.

Pros

- Ideal for sloped landscapes.
- Helps control erosion.
- Less material required than fully enclosed raised beds.

Cons

- May limit access or mobility.
- Requires strategic design to ensure stability.

The best garden bed for you depends on your space, budget, and goals. Some gardeners may prefer a combination of methods—for example, raised container beds for vegetables, in-ground beds for perennials, and pocket gardens for edible flowers.

By selecting the method that best suits your landscape, you can create a productive, low-maintenance vegetable garden that enhances your soil and supports biodiversity for years to come!

Winter Garden Maintenance

As temperatures drop and plants enter dormancy, many gardeners shift into fall cleanup mode—removing dried flower stalks, trimming perennials, and raking away leaves. However, conventional fall landscaping practices can do more harm than good, stripping the garden of essential nutrients, habitat, and winter food sources. Instead of viewing dried vegetation as "dead" material to be discarded, a regenerative approach allows nature to recycle its resources while also supporting biodiversity.

By making small shifts in how we manage plant material in autumn and winter, we can improve soil health, create habitat, and even enhance the beauty of the winter garden. Here are four key reasons to rethink your fall landscaping habits and simple techniques for a permaculture-inspired approach to seasonal plant management.

KEEP YARD TRIMMINGS ON-SITE TO FEED THE SOIL

Instead of bagging up plant debris or sending yard waste to a landfill, return those nutrients to the soil by leaving trimmings in your planting beds. When you cut back perennials, grasses, or spent stems, chop them into 3-to-5-inch (8 to 13 cm) pieces and let them decompose in place naturally. This mimics nature's natural recycling process, where organic material breaks down, enriching the soil with valuable nutrients.

How to Recycle Plant Material in Your Garden

- **Chop and drop.** Cut stems and foliage into small sections and drop them in garden beds to act as natural mulch.
- **Pile and hide.** If you prefer a neater look, stack plant material behind shrubs or in less visible areas to decompose over time.
- **Leave partial stalks intact.** Keep 4 to 5 inches (10 to 13 cm) of aboveground stalks standing to create habitat for overwintering insects like solitary bees.

Left Instead of viewing dried vegetation as "dead" material to be discarded, leaving the dried thatch over winter provides essential food sources and habitat for songbirds and beneficial insects during the cold months.

Opposite Maximize space by incorporating vegetables and edible flowers into existing ornamental beds. If you have limited sun or space, pocket gardening is an easy way to integrate food production into your existing landscape.

By following these methods, you retain organic matter on-site, which builds soil fertility, reduces the need for external fertilizers, and improves moisture retention in garden beds.

DRIED FLOWER HEADS ADD WINTER BEAUTY AND STRUCTURE

A common misconception is that a winter garden is dull and lifeless but leaving dried flower heads and seed pods intact can add architectural interest and texture to your landscape.

How Dried Flowers Enhance the Winter Garden

- **Seed heads create visual contrast.** The sculptural shapes of plants like coneflowers (*Echinacea* spp.), black-eyed Susans (*Rudbeckia* spp.), and ornamental grasses provide movement and structure.
- **Frost and snow transform dried plants.** Ice crystals settle beautifully on persistent stems, adding seasonal charm.
- **Shapes and patterns stand out.** Without the distraction of summer foliage, dried flowers create unique silhouettes against the winter sky.

Instead of cutting everything down at season's end, allow perennials to stand through winter. Not only will this enhance your landscape's aesthetic appeal, but it also offers ecological benefits that support wildlife and biodiversity.

SEED HEADS PROVIDE A WINTER FOOD SOURCE FOR BIRDS AND WILDLIFE

More important than their visual appeal, dried seed heads play a crucial role in supporting wildlife through the colder months. Many birds that rely on insects during warmer seasons shift to eating seeds in winter, making plants like echinacea, sunflowers (*Helianthus* spp.), and grasses a lifeline when food sources are scarce.

Ways to Support Wildlife with Dried Plants

Leave Seed Heads on Flowers

Perennials like coneflowers (*Echinacea* spp.) and asters (*Symphyotrichum* spp.) provide an essential winter food source for finches, sparrows, and other seed-eating birds.

Let Ornamental Grasses Stand

Their dried seed stalks are eaten by birds while also offering shelter from harsh weather.

Consider Small Brush Piles

Gather plant stems and place them in an undisturbed corner of the yard to create habitat for small mammals and ground-dwelling birds.

By resisting the urge to deadhead plants in fall, your landscape will actively support local wildlife while reducing the need for artificial bird feeders.

DRIED VEGETATION PROVIDES SHELTER FOR BENEFICIAL INSECTS AND SMALL CREATURES

Aside from feeding birds, dried plant stalks and grasses also provide critical winter shelter for beneficial insects, reptiles, and mammals. Many overwintering pollinators and predatory insects—including butterflies, ladybugs, and solitary bees—use stems, fallen leaves, and hollow plant stalks as their winter refuge. If all plant material is removed from the landscape, these species lose their natural shelter, which reduces pollinator populations in spring.

How to Protect Overwintering Insects and Wildlife

Leave Standing Plant Stems Until Early Spring

Butterflies and moths lay eggs on dried vegetation while native bees burrow inside hollow stalks.

Keep a Portion of Dried Grasses Intact

Frogs, lizards, and small reptiles use thick plant thatch for winter cover.

Delay Cutting Back Meadows and Pollinator Gardens

In temperate climates, wait until the first signs of spring before removing dried vegetation to avoid disturbing hibernating insects.

If tidiness is a concern, cut back only select areas while leaving sections of the garden untouched for wildlife. Consider a "messy corner" approach, where a portion of your yard remains undisturbed for habitat purposes.

By rethinking traditional fall cleanup habits, you create a winter friendly landscape that enhances biodiversity, retains nutrients, and provides critical food and shelter for wildlife.

Instead of removing dried flowers, grasses, and leaf litter, allow nature to take its course. Your garden will remain visually interesting, support birds and beneficial insects, and set the stage for healthier plant growth in spring.

Simple Shifts to Make Your Garden More Regenerative in Fall and Winter

- Leave seed heads and dried flowers intact for winter interest and wildlife food.
- Cut and drop perennial trimmings into beds to return nutrients to the soil.
- Keep 4 to 5 inches (10 to 13 cm) of plant stalks standing to create insect habitat.
- Avoid cutting back pollinator-friendly gardens until early spring.

Now that we've covered the fundamental principles and techniques of regenerative landscaping—water management, soil health, and plant biodiversity—it's time to see these concepts in action. Section 4 features real-world case studies of permaculture landscapes, demonstrating how different approaches can be tailored to fit a variety of spaces and lifestyles.

Top Choosing plants with winter nectar for pollinators is a great way to support pollinators year-round, especially as we experience warmer winters. Witchhazel (*Hamamelis virginiana*) provides bursts of yellow blossoms in mid-winter.

Bottom Alpine strawberry (*Fragaria vesca*) is a low-growing, perennial ground cover that spreads through runners, forming a dense mat of lush, green foliage while producing small, flavorful berries. Thriving in partial shade to full sun, it helps suppress weeds, stabilizes soil, and attracts pollinators, making it an excellent choice for edible and ecological landscaping.

Blueberry (*Vaccinium virgatum*) ripening in a client's yard. Blueberry is a versatile fruiting shrub that produces prolifically while thriving in many types of light and soil conditions.

Section 4

Property Case Studies

As we've discussed, at its essence, permaculture seeks to integrate human needs and activities with the needs and yields of natural ecosystems. Regardless of your landscape terrain, location, soil profile, light availability, climate, or goals, you can "do permaculture" in any context. The actual plants you install and techniques you implement will depend on the specific conditions of your site, but all permaculture gardens will draw upon the Three Pillars of a Regenerative Landscape. In the coming pages, I will show you actual residential projects Shades of Green has designed and installed, each of which addresses a different design challenge and client goal, while implementing permaculture principles using site-appropriate techniques.

When we ground our landscape designs in the three core pillars of regeneration, we create spaces that nourish both people and the planet.

Let's explore real-world projects that have embodied these principles, showing how they can be adapted to any landscape, from tiny shady slopes to sunny expanses in upscale neighborhoods to productive homesteads. Each case study reveals how intentional design choices, rooted in nature's own intelligence, have restored ecological function, increased resilience, solved common drainage issues, and provided abundance in ways that are both beautiful and practical.

Theory and techniques are powerful, but seeing real-world examples is what truly brings permaculture to life. It's one thing to read about water harvesting, soil regeneration, and biodiverse planting strategies—it's another to witness how these principles transform ordinary yards into thriving, resilient ecosystems.

"You never change things by fighting the existing reality. To change something, build a new model that makes the existing model obsolete."

BUCKMINSTER FULLER

From Flooded Lawn to Thriving Garden: A Multifunctional and Compact Urban Space

Even tiny spaces, if intentionally designed, can grow food, manage water, and have room to gather. Here's how we transformed a compact backyard into a productive urban garden.

THE SITE

Layout and Topography

This backyard is about 1,500 square feet (139 sq m) and serves as the main thoroughfare from the homeowners' parking area to their back deck, the primary entrance to their home.

Problem Areas and Site Challenges

The backyard sits at a low point in the watershed, collecting runoff from neighboring properties. Water flows in from the neighboring yard, which slopes toward this space, and additional runoff comes from a large neighboring roof. Altogether, about 6,000 square feet (557 sq m) of drainage area funnels water into this small yard, causing standing water and excessive moisture.

Existing Conditions

Before the transformation, the backyard featured a patchy, underutilized lawn with compacted soil and struggling ornamental plants. Poor drainage led to frequent flooding, making the space difficult to use and leaving many plants diseased or failing. Since this space also functioned as the main pathway between the parking area and the back entrance, it needed to be durable and practical, not just pretty.

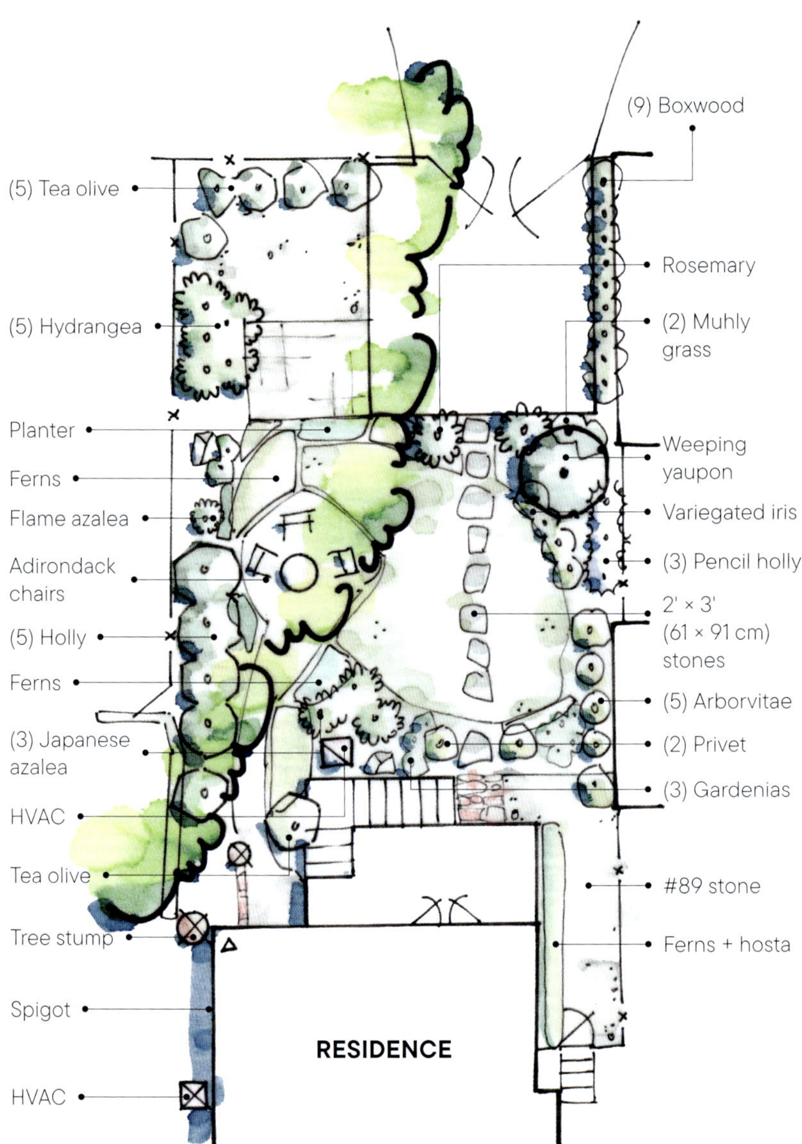

The site originally struggled with disease-prone ornamentals and poor drainage, as seen in this site analysis map.

Before

Above This client wanted to transform their unused, high-maintenance lawn space into something more productive, beautiful, and useful.

Top right Because of water issues and too much shade, the lawn was consistently struggling to survive.

Right The neighbor's downspouts drain into this small backyard, which was causing a lot of flooding and erosion as the land struggled to manage the incoming volume of water.

CASE STUDY #1: A TINY URBAN OASIS

THE GOAL

The homeowners envisioned a backyard that did more than just sit there—it needed to work for them and serve multiple functions. They wanted a space that could handle heavy rains without flooding, making it easier to navigate year-round. Since their backyard serves as the main walkway from the parking area to their home, it had to be both functional and inviting.

Beyond practicality, they dreamed of a lush, productive garden where they could grow an abundance of food, herbs, and flowers. They also wanted to create a space where they could truly enjoy being outside—gathering around the fire pit on cool evenings, tending plants in the reclaimed-window greenhouse, and relaxing on their newly renovated, tiered deck with comfortable outdoor seating. By replacing the struggling lawn with a vibrant, biodiverse landscape, they aimed to build a backyard that was not only beautiful and ecologically beneficial but also a welcoming retreat for daily life.

THE SOLUTION

Water: Fixing Drainage and Infiltration

The first priority was managing water. Instead of letting rainwater pool and create a muddy mess, we designed the space so it could absorb and filter water naturally. We built permeable pathways using a base of crushed recycled concrete, topped with repurposed slate chips. These paths allow

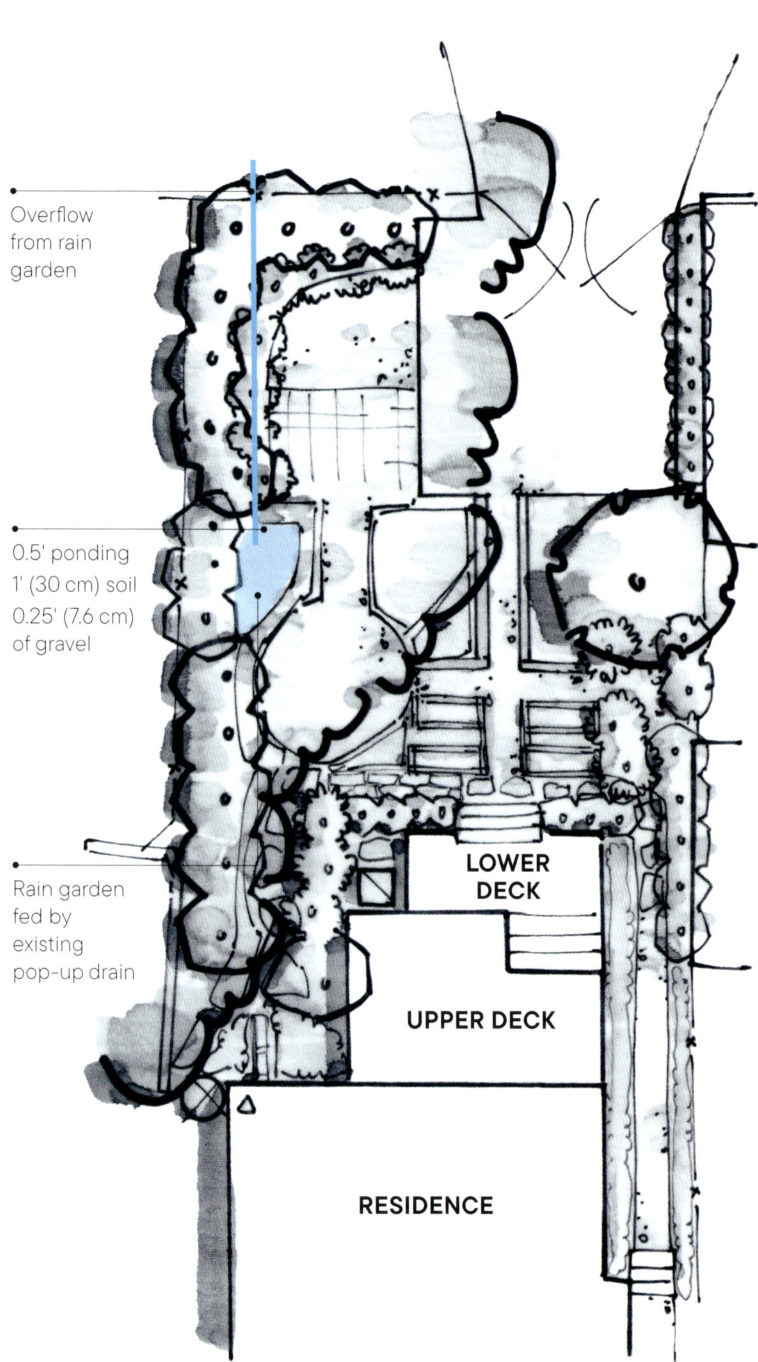

Routing the roof water into a rain garden then overflowing the backyard helped mitigate the drainage issues.

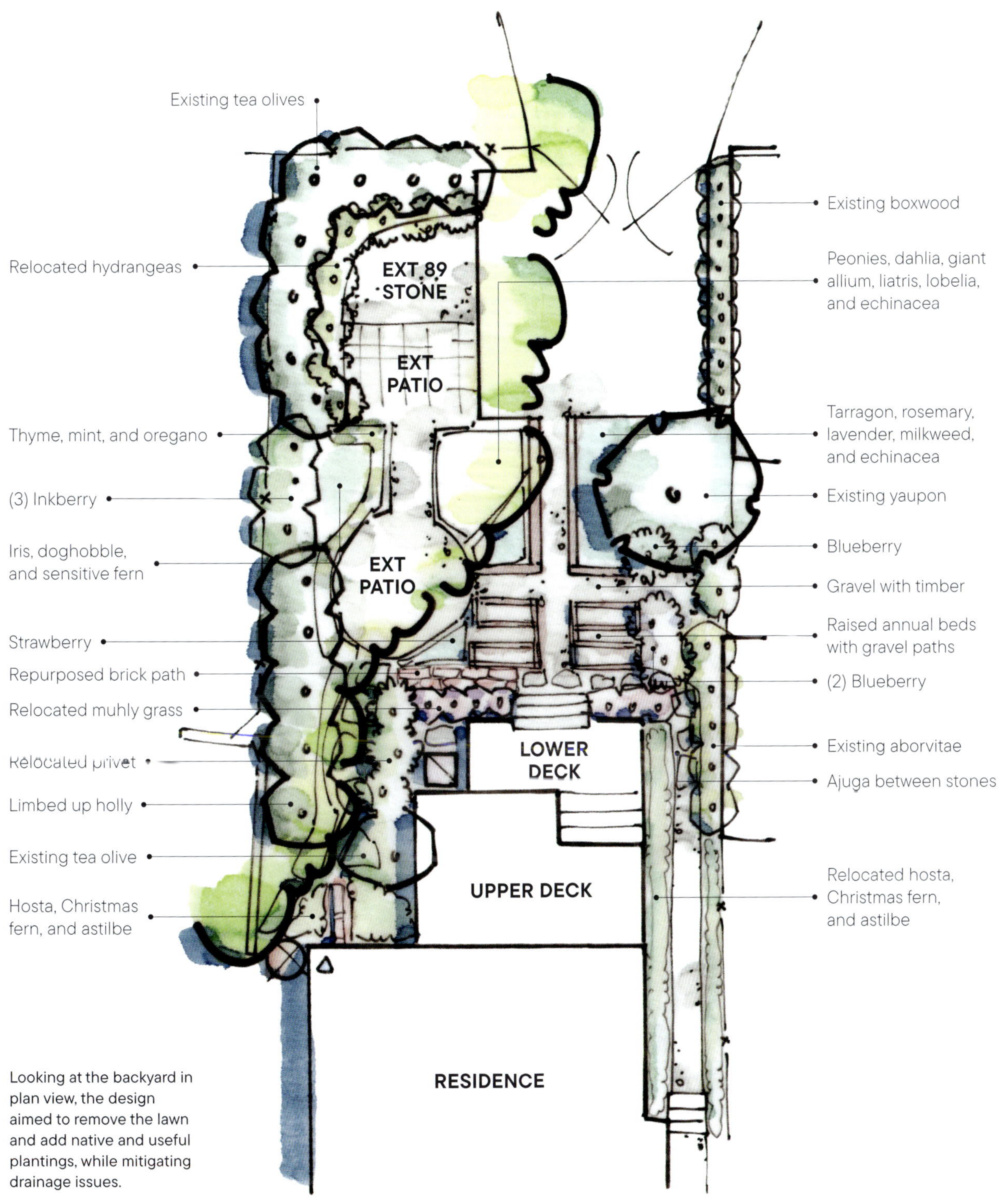

Looking at the backyard in plan view, the design aimed to remove the lawn and add native and useful plantings, while mitigating drainage issues.

After

Opposite, top left A rain garden blends into the surrounding garden beds and provides a low spot for water to infiltrate before overflowing toward the back alleyway.

Opposite, top right The rain garden, in the right of the photo, is planted with seasonally interesting native plants, such as doghobble (*Leucothoe fontanesiana*) and cardinal flower (*Lobelia cardinalis*).

Opposite, bottom left View from the back deck, looking over the herb and vegetable garden. A small rain garden for runoff is situated between the firepit and greenhouse.

Below The gardens produce an abundance of food, culinary and medicinal herbs, and cut flowers, all in under 1,500 square feet (139 sq m).

rainwater to sink into the soil instead of running off, preventing erosion and puddling.

To handle even more water, we installed a rain garden, planted with doghobble (*Leucothoe fontanesiana*), cardinal flower (*Lobelia cardinalis*), mint (*Mentha* spp.), and blazing star (*Liatris spicata*). These native plants thrive in wet conditions, slow down runoff, and provide habitat for pollinators.

Soil: Building Fertility for a Productive Garden

Once the drainage issue was under control, we turned our attention to the soil. We removed the compacted lawn and built raised garden beds, filling them with rich, organic soil to support healthy plant growth. These beds now produce a mix of food and flowers, including tomatoes (*Solanum lycopersicum*), basil (*Ocimum basilicum*), strawberries (*Fragaria × ananassa*), oregano (*Origanum vulgare*), and dahlias (*Dahlia* spp.).

Plants: A Thriving and Functional Landscape

Instead of struggling grass, the backyard now overflows with a mix of edible and native plants. Hydrangea (*Hydrangea quercifolia*), lavender (*Lavandula angustifolia*), and echinacea (*Echinacea purpurea*) add color and attract pollinators. Weeping yaupon holly (*Ilex vomitoria* 'Pendula'), a native tea-producing evergreen, provides privacy and year-round greenery. Grasses like little bluestem (*Schizachyrium scoparium*) and pink muhly grass (*Muhlenbergia capillaris*) add texture and movement while supporting local wildlife.

A BACKYARD THAT WORKS

Now, this once-wet and wasted space is a thriving, functional urban garden. The pathways make daily movement easy, the plants soak up excess water, and the whole space supports biodiversity while producing fresh food and flowers. Thoughtful design transformed this backyard into a beautiful, productive, and resilient landscape that works for both people and nature.

Opposite, top left Managing overland water required a combination of earthworks and dense ground cover.

Opposite, top right Mixing aromatic herbs with veggies and cut flowers helps to maximize the space and reduce pest pressure.

Opposite, bottom Cut flowers, such as this *Dahlia* spp., add beauty that can be brought inside, helping the client meet the goal of connecting the interior and exterior spaces and add lots of color in and around the exterior gathering spaces.

Right, top Potted veggies and herbs help take advantage of the limited sunlight.

Right, bottom The view looking from the driveway to the back of the house, with lots of outdoor living spaces, vegetables, herbs, and cut flowers.

DIY RECLAIMED-WINDOW GREENHOUSE: A BEAUTIFUL, FUNCTIONAL BACKYARD RETREAT

A greenhouse is one of the most versatile additions you can make to your backyard. It extends the growing season, provides shelter for plants, and even doubles as a cozy outdoor space for dining or relaxing. But instead of buying an expensive prefabricated model, why not build one using salvaged materials?

This DIY reclaimed-window greenhouse is an affordable, sustainable, and beautiful way to repurpose old windows into a functional garden structure. With a little creativity, some elbow grease, and a few key materials, you can create a space that blends seamlessly into your landscape while serving multiple purposes year-round.

WHY USE RECLAIMED WINDOWS?

Using salvaged windows gives your greenhouse a unique, vintage charm while reducing waste and saving money. Instead of sending old windows to the landfill, you're giving them a second life. Plus, since many old windows are made with high-quality wood and glass, they add a level of craftsmanship that's hard to replicate with modern materials.

Optional: If you want to insulate the greenhouse for colder months, consider adding weather stripping or caulking between the window seams.

INSTRUCTIONS

Step 1: Gather and Prepare Materials

Start by sourcing reclaimed windows. Check salvage yards, thrift stores, or online marketplaces like Craigslist and Facebook Marketplace. If you can't find matching sizes, arrange the windows in a way that balances aesthetics with structural integrity. To create a cohesive look, paint all the windows in the same color before assembling. A soft green, white, or black works well for a classic greenhouse feel.

Step 2: Build the Foundation

A solid, level base is key to a long-lasting greenhouse. For a rustic and low-maintenance option, lay a reclaimed brick floor over a bed of leveled sand or gravel. Dry-stacking the bricks allows

A 5 foot by 8 foot (1.5 × 2.4 m) greenhouse made of reclaimed windows with flexible outdoor dining space.

Materials You'll Need

- Reclaimed windows (matching sizes make things easier, but a mix can add character)
- 2 × 4-inch (38 × 89 mm) or 2 × 6-inch (38 × 140 mm) lumber (for framing)
- Reclaimed bricks (for flooring)
- Wood screws, nails, and brackets (for assembling the frame)
- Exterior paint (to create a uniform look)
- Hinges and a door handle (for the entryway)
- Gravel or sand (for leveling the foundation)
- Clear polycarbonate or salvaged glass panels (for the roof)

water to drain while giving the space a charming, old-world look.

Step 3: Construct the Frame

1. Build a simple rectangular frame using 2 × 4-inch (38 × 89 mm) or 2 × 6-inch (38 × 140 mm) lumber. This will serve as the main support for attaching the windows.
2. Secure the windows to the frame with wood screws or brackets, making sure they're tightly fitted to prevent drafts.
3. Install a salvaged door or build one using a larger window attached to hinges.

Step 4: Add the Roof

For the roof, you can use clear polycarbonate panels for durability or repurpose old glass panels if they're available. Attach them securely to the frame, ensuring they have a slight slope for rainwater runoff.

HOW TO USE YOUR GREENHOUSE

Once built, your reclaimed-window greenhouse can serve multiple purposes throughout the year:

The small greenhouse adds a backdrop to the outdoor gathering space and creates some privacy between the back alleyway used to access adjacent properties and the main areas in the backyard.

- Start seedlings early: Protect young plants from frost and extend your growing season.
- Grow herbs year-round: Keep fresh herbs like rosemary, thyme, and basil thriving in cooler months.
- Outdoor dining space: Set up a small table and chairs to enjoy meals in a cozy, sunlit spot.
- Plant storage and propagation: Keep delicate plants safe during winter or propagate new cuttings.
- A peaceful retreat: Use it as a reading nook, a workspace, or simply a place to enjoy your garden.

A BEAUTIFUL, FUNCTIONAL ADDITION TO ANY BACKYARD

Building a reclaimed-window greenhouse is a rewarding project that combines sustainability, function, and beauty. With a little effort, you can create a space that not only helps your plants thrive but also becomes one of your favorite spots in the garden. Whether you use it for growing, gathering, or simply unwinding, this greenhouse is proof that salvaged materials can create something truly special.

A Thriving Herbalist's Garden

I want to walk you through a project we did for a well-known Atlanta-based herbalist. We had the pleasure of working with the client to design a space where they could grow an abundance of food and medicinal plants for their family and herbal practice.

What made this project particularly inspiring was the client's deep involvement in selecting the plants, their ongoing interaction with the space as a living ecosystem, and their commitment to stewarding not only their own property but also the empty land next door and the stream behind their home. They saw this work as an act of respect and reciprocity—caring for the land while allowing it to support them in return.

THE SITE

Layout and Topography

This property is situated mostly in a floodplain, with a creek running behind the client's house on an adjacent, city-owned property. A relatively flat site, during heavy rains the creek behind the house rises, and the backyard floods. The front yard is compact, yet sunny—ideal for growing vegetables and full-sun herbs. The backyard is a little shadier and moister, with pockets of sunlight from the dappled canopy above. Additionally, the bulk of the apothecary's row production and other family needs, such as entertaining, raising chickens, and accessing the trails along the creek, would inform the design of a multifunctional space.

Above Herbs climb up a trellis at this client's home apothecary garden.

Opposite, left The front yard is compact, yet sunny—ideal for growing vegetables and full-sun herbs.

Opposite, right The backyard is a little shadier and moister, with pockets of sunlight from the dappled canopy above.

Before

Problem Areas and Site Challenges

While the site had great sun exposure, there were some drainage challenges. The home's downspouts were emptying at the foundation, creating moisture issues in the crawlspace. Additionally, overland runoff was eroding sections of the yard, exposing the subsoil and making it difficult to establish healthy plantings.

Existing conditions

Like many properties in Atlanta, the client's space originally featured an expanse of underutilized full-sun lawn in the front yard, which presented an incredible opportunity for growing a diverse range of medicinal and edible plants. Knowing that we were working in both a floodplain and in a diversity of microclimates—the forest edge, full-sun front yard, moist creekside—helped us craft a plant palette spanning a wide

CASE STUDY #2: AN HERBALIST'S HOME APOTHECARY

Mapping the existing conditions helps get a sense of the site. Knowing that we were working in both a floodplain and in a diversity of microclimates—the forest edge, full-sun front yard, moist creekside—helped us craft a plant palette spanning a wide range.

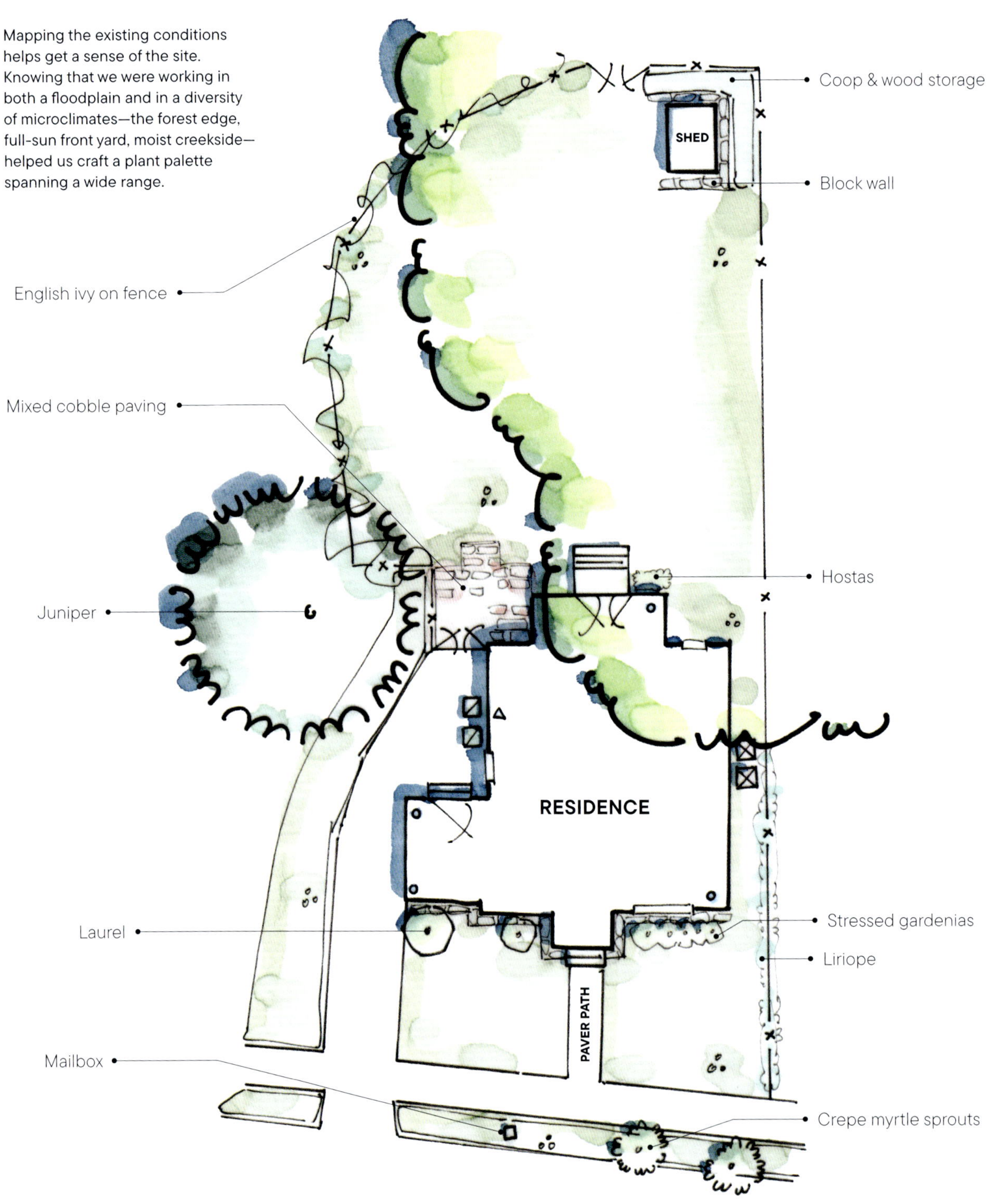

Top Like many properties in Atlanta, the client's space originally featured an expanse of underutilized full-sun lawn in the front yard, which presented an incredible opportunity for growing a diverse range of medicinal and edible plants.

Bottom Growing a variety of medicinal and edible plants was the highest priority for this client.

range. In addition, all of the hardscaping and paths were designed to handle floodwaters without washing out.

THE GOAL

The client's vision extended far beyond simply growing food and medicine—they wanted to create a space that functioned as both a productive garden and a regenerative ecosystem. First and foremost, they needed a way to manage stormwater effectively to prevent erosion and moisture buildup around their home. At the same time, they wanted to cultivate a variety of growing conditions that could support the diverse range of medicinal and edible plants essential to their herbal practice.

Beyond the practical concerns, the client really valued their land as part of a larger ecological system. Restoring soil health was a priority, but they also wanted their garden to contribute to biodiversity, supporting pollinators and beneficial wildlife while integrating native species that would thrive in the local environment. Stewardship was at the heart of their approach—not just for their own property, but for the adjacent land and the stream that ran behind it. For them, tending this garden was an act of reciprocity, a way to care for the land while allowing it to care for them in return.

Our goal was to transform the space into a functional and abundant garden with a variety of microclimates to accommodate the wide range of medicinal plants the client wanted to grow. At the same time, we needed to integrate a stormwater management system that would protect their home, restore soil health, and work harmoniously with their vision for the land as well as the existing conditions.

THE SOLUTION

Water: Managing Stormwater with Rain Gardens and Swales

We addressed the site's drainage challenges by directing the home's downspouts into rain gardens in the front yard. These gardens temporarily hold rainwater, allowing the rain to sink into the soil instead of pooling near the foundation. The plants in these rain gardens need to withstand both flood and drought conditions, as they experience cycles of total saturation and complete dryness.

In the backyard, we implemented a similar approach, directing runoff into rain gardens that overflow into a contoured swale running parallel to a creek-access path. The design allowed water to sheet-flow evenly toward the creek rather than rushing downhill in concentrated runnels, which helps hydrate the soil while preventing further erosion. By slowing and spreading water across the landscape, we created a system that stabilized the soil and increased resilience against both drought and heavy rains.

Top Anise hyssop (*Agastache foeniculum*) serves many functions in the landscape. Its fragrant, nectar-rich flowers provide medicinal and culinary benefits, drought tolerance, and deer resistance.

Bottom Wild bergamot (*Monarda fistulosa*) is a drought-tolerant native perennial that attracts bees, butterflies, and hummingbirds while supporting biodiversity with its long-lasting lavender blooms. Its aromatic foliage has antimicrobial properties, historically used in herbal medicine, and it helps stabilize soil while resisting deer and rabbits.

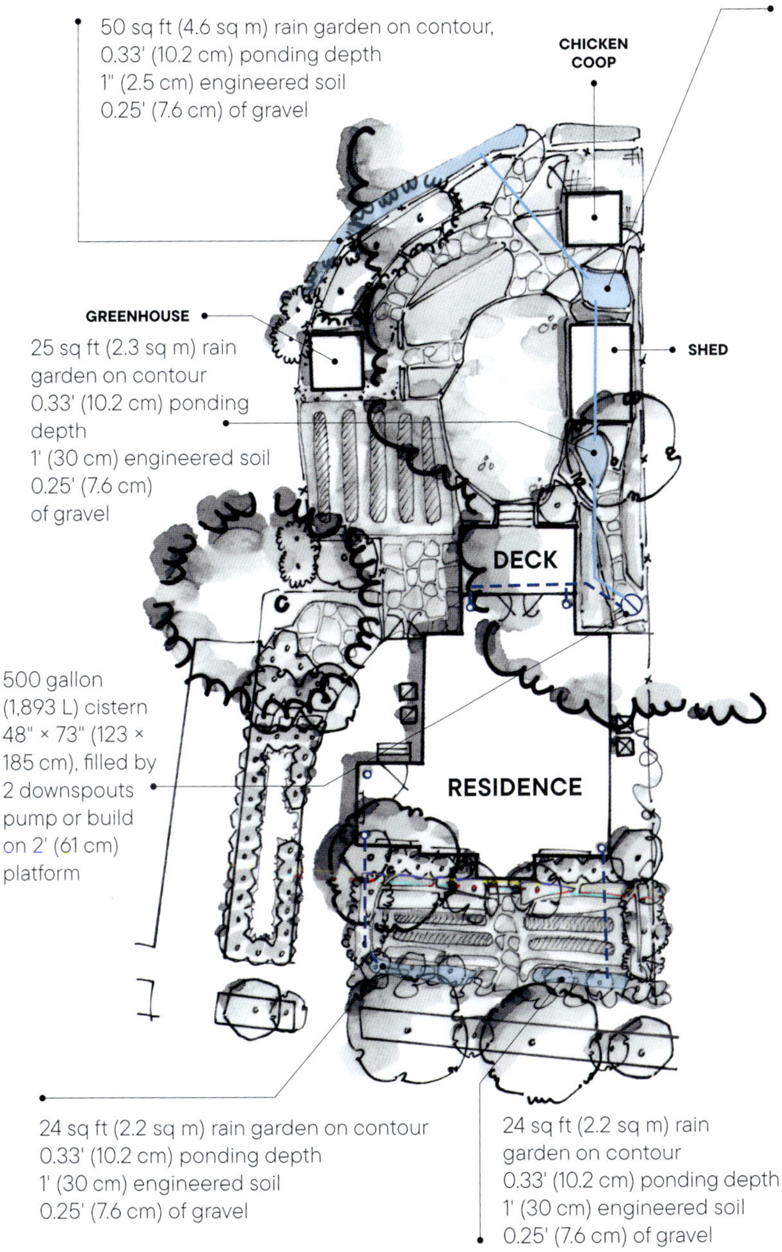

50 sq ft (4.6 sq m) rain garden on contour,
0.33' (10.2 cm) ponding depth
1" (2.5 cm) engineered soil
0.25' (7.6 cm) of gravel

CHICKEN COOP

25 sq ft (2.3 sq m) rain garden on contour
0.33' (10.2 cm) ponding depth
1' (30 cm) engineered soil
0.25' (7.6 cm) of gravel

GREENHOUSE

25 sq ft (2.3 sq m) rain garden on contour
0.33' (10.2 cm) ponding depth
1' (30 cm) engineered soil
0.25' (7.6 cm) of gravel

SHED

DECK

500 gallon (1,893 L) cistern
48" × 73" (123 × 185 cm), filled by 2 downspouts
pump or build on 2' (61 cm) platform

RESIDENCE

24 sq ft (2.2 sq m) rain garden on contour
0.33' (10.2 cm) ponding depth
1' (30 cm) engineered soil
0.25' (7.6 cm) of gravel

24 sq ft (2.2 sq m) rain garden on contour
0.33' (10.2 cm) ponding depth
1' (30 cm) engineered soil
0.25' (7.6 cm) of gravel

Soil: Restoring Fertility with *Hügelkultur*

With an abundance of downed branches and wood on-site, we saw an opportunity to build *hügelkultur* beds—raised garden beds filled with logs, branches, and organic matter. These beds serve a dual purpose: They restore degraded soil and provide long-term moisture retention for crops.

Since many of these beds were placed in areas affected by erosion, they also helped slow the movement of water across the site, allowing it to infiltrate and recharge the soil. Over time, the decomposing wood acts like a sponge, holding moisture and slowly releasing it to plant roots, reducing the need for irrigation. These beds now support an array of herbs, vegetables, and large row crops, which are thriving in the naturally enriched soil.

Above The design sought to channel rainwater into the landscape through rain gardens and *hügelkultur* beds, allowing the water to sink into the soil and passively irrigate all the planting beds throughout the site.

Right Downspout filters keep mosquito larva from entering the feedlines and seal the downspouts so the water can route via gravity to rain gardens in both the front and backyard.

After

Top The in-ground *hügelkultur* beds absorb overland water in the front yard and minimize the need for irrigation.

Bottom What was once a barren grassy lawn, struggling to grow anything, is now a vibrant medicinal and edible oasis buzzing with life.

Top Rain gardens just after planting in the dappled understory of the backyard.

Bottom A permeable flagstone path meanders through a woodland edge planting bed.

Plants: A Biodiverse Herbal Sanctuary

As an herbalist, the client wanted to fill their garden with as many useful plants as possible, making use of every niche and microclimate on their property. Their plant selections reflected both their personal practice and their commitment to ecological restoration.

We designed the garden to include a mix of native meadow species such as echinacea (*Echinacea purpurea*) and goldenrod (*Solidago* spp.) to attract pollinators and provide medicinal benefits. In the shadier woodland edges, we introduced bloodroot (*Sanguinaria canadensis*) and cohosh (*Actaea racemosa*), both valuable medicinal plants native to the southeastern US. Showier flowering plants like roses (*Rosa* spp.) and vitex (*Vitex agnus-castus*) add beauty while serving as powerful allies in their apothecary.

By incorporating a diverse mix of native, medicinal, and edible plants, we created a landscape that not only provides for the client's herbal practice but also enriches the surrounding ecosystem.

A GARDEN ROOTED IN RECIPROCITY

This project went beyond just planting a garden—it was about deepening a relationship with the land. By restoring soil health, slowing and filtering rainwater, and filling the space with life-sustaining plants, we helped create a garden that supports both the people who tend it and the environment around it.

The client's daily interactions with the space—harvesting plants, tending the soil, and stewarding the land beyond their property—bring the garden to life in a way that is both deeply personal and ecologically impactful. This thriving herbal sanctuary now stands as a testament to the power of thoughtful, intentional land stewardship, blending beauty, productivity, and regeneration into one harmonious landscape.

Top The front beds now support an array of herbs, vegetables, and large row crops, thriving in the naturally enriched soil.

Bottom Black-eyed Susan (*Rudbeckia hirta*) is a hardy, drought-tolerant wildflower that provides nectar for pollinators, supports native bees, and adds long-lasting golden blooms to the landscape. Its seeds also feed birds in the fall, making it a valuable plant for wildlife-friendly gardens.

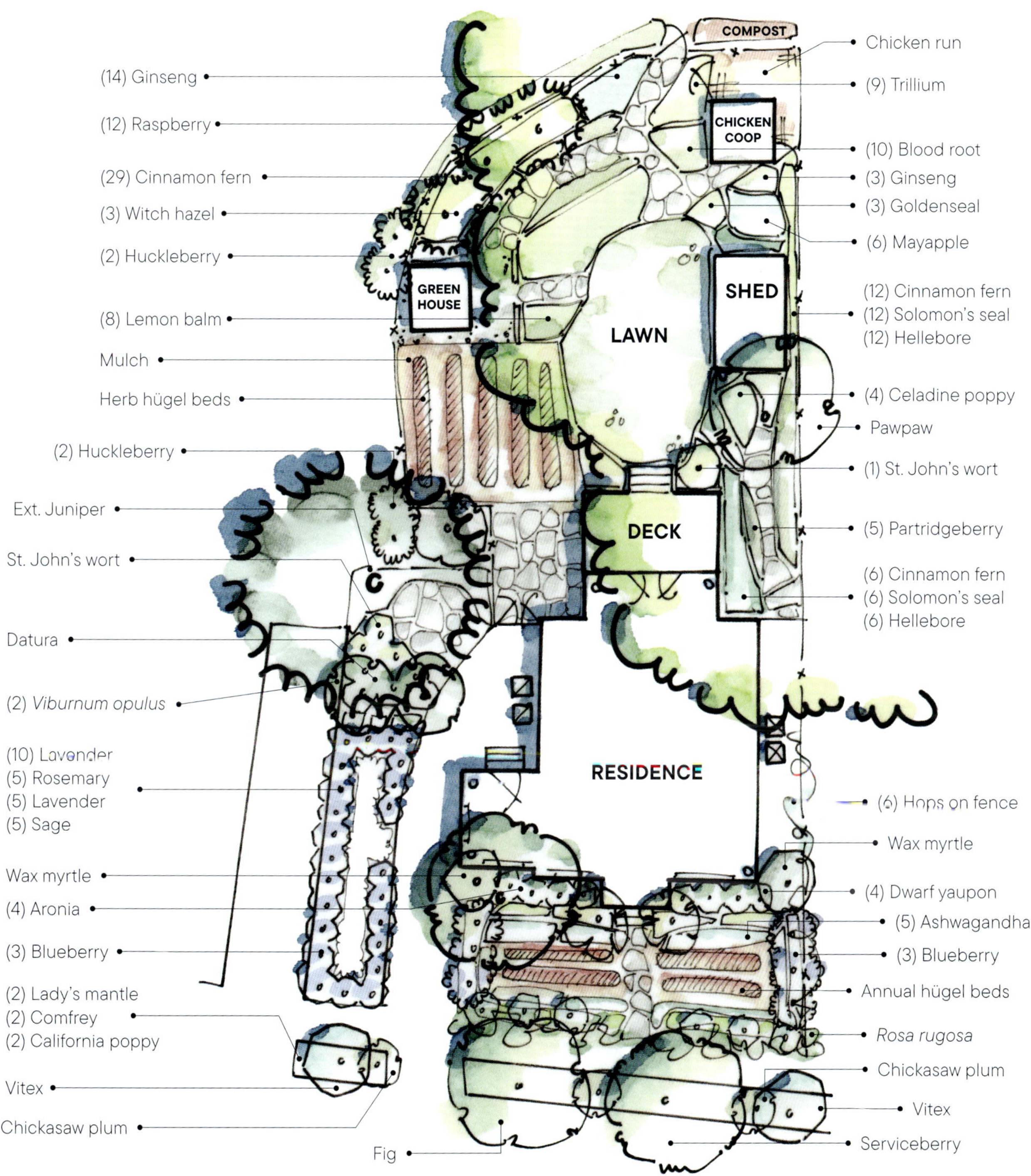

The planting design seeks to fill the garden with as many useful plants as possible, making use of every niche and microclimate on their property.

DIY A-FRAME TRELLIS USING NATURAL MATERIALS

If you're looking for a simple, sturdy, and eco-friendly way to support climbing plants like cucumbers, pole beans, peas, or vining flowers, an A-frame trellis is a great option. It maximizes vertical growing space, improves airflow, and makes harvesting easier. Best of all, you can build one using natural and salvaged materials like bamboo and twine.

A homemade A-frame trellis straddles a vegetable *hügelkultur* bed. The A-frame design maximizes growing space and is easy to create using reclaimed materials you might have lying (or growing) around.

WHY AN A-FRAME TRELLIS?

This type of trellis is perfect for small-space gardens because it allows plants to grow upward instead of sprawling across the ground. Using an A-frame design means:

- **Better airflow:** Reduces the chance of mildew and disease.
- **More growing space:** Keeps vines off the ground so other crops can be planted underneath.
- **Stronger support:** The angled design can handle the weight of heavy fruiting plants like cucumbers.
- **Sustainable and budget-friendly:** Can be made from natural, biodegradable materials.

Materials You'll Need

- → Bamboo poles (at least 6 feet [1.8 m] long, about six to eight pieces per trellis)
- → Twine or natural jute rope
- → Garden stakes or extra bamboo for stability
- → Scissors or garden shears

Optional: If bamboo isn't available, you can use small tree branches, untreated wooden stakes, or even repurposed pallet wood.

INSTRUCTIONS

Step 1: Choose Your Location

Pick a sunny spot in your garden where your climbing plants will thrive. If you're placing the trellis near a house or fence, make sure there's enough space for airflow.

Step 2: Build the A-Frame

1. Lay out your bamboo poles. Arrange them into two equal sets (three or four per side, depending on the width you want).
2. Form the A-shape. Lean two poles together at the top to create an inverted *V*.
3. Secure the top. Use twine or jute rope to tie the poles together tightly. Wrap the twine several times, crossing in an *X* pattern for extra strength.
4. Repeat. Make a second A-frame structure using another pair of bamboo poles.

Step 3: Connect the Two Sides

1. Place the A-frames in the garden. Position them 2 to 3 feet (61 to 91 cm) apart, depending on the length of your trellis.
2. Tie a horizontal bamboo pole across the top. This stabilizes the structure and keeps it from collapsing outward.
3. Add support crosspieces. Attach additional horizontal bamboo poles or sturdy twine between the two frames to give climbing plants something to grab onto.

Step 4: Create the Climbing Grid

1. Weave twine between the poles. Tie twine horizontally across the frame every 6 inches (15 cm), securing each end tightly.
2. Add vertical supports. Tie additional pieces of twine from the top to the bottom, creating a grid or net-like structure for tendrils to climb.

Tips for Success

- For extra stability, push the bamboo poles a few inches into the ground before tying them together.
- Use biodegradable twine instead of plastic netting to keep the trellis eco-friendly.
- Grow compatible crops underneath. Consider planting shade-tolerant vegetables like lettuce or radishes beneath the trellis to maximize your space.
- Harvest frequently. Keeping fruit and vegetables off the ground reduces pests and improves airflow.

A Simple, Sustainable Solution for Your Garden

Building your own A-frame trellis is an easy way to add structure to your garden while using natural and sustainable materials. Whether you're growing cucumbers, beans, or flowering vines, this trellis design will help you get the most out of your space while keeping your plants healthy. Plus, it blends beautifully into any garden setting, adding a rustic, handmade touch to your growing space.

Pocket-Sized Productive Shade Garden

As you've seen so far in the previous sections, I've narrowed regenerative landscaping down to three primary areas of influence for residential land management: restoring the water cycle, building soil fertility, and cultivating diverse plant communities that protect biodiversity while growing food, medicine, and pollinator habitat.

To illustrate this approach, I want to highlight a project I designed and installed with my partner on our mostly shaded, sloped, tenth of an acre (less than 6,000 square feet [557 sq m]) lot—our own home. I chose this project because the patterns translate to much of our residential clientele in this sub/urban region.

Even small, shaded spaces can be transformed into thriving, productive gardens. This project was a perfect example of how thoughtful design can turn an underutilized, overgrown backyard into a lush, food-producing ecosystem while restoring soil health, managing stormwater, and supporting biodiversity.

THE SITE

My property sits on a north-facing slope within a forested landscape dominated by pines (*Pinus* spp.), red oak (*Quercus rubra*), sweetgum (*Liquidambar styraciflua*), and red maple (*Acer rubrum*). The highest point on the property is the backyard, which also happens to be the only flat and sunny area, receiving about five hours of direct sunlight at the summer solstice.

Above My daughter, Zephyr Dove, running through our garden.

Opposite, left Flame azalea (*Rhododendron calendulaceum*) is native to the Appalachian Mountains of the eastern United States, thriving in woodlands, mountain slopes, and open meadows, where it provides nectar for pollinators like bees, butterflies, and hummingbirds.

Opposite, right My family enjoys the bounty of our landscape. Seeing our daughter graze herbs and connect with the plants and animals that we share our yard with is a joy that drives my passion for this work.

Like many urban landscapes, this site faced serious erosion issues from overdevelopment, loss of topsoil, and mismanaged water. Years of unchecked runoff had stripped away topsoil, leaving behind compacted clay with very little organic matter. The compaction also contributed to poor drainage, leading to water pooling around the foundation of the house. Meanwhile, the north slope was completely overtaken by English ivy (*Hedera helix*), a fast-spreading aggressive species notorious for suppressing native plants and attracting mosquitoes.

THE GOAL

Despite its challenges, the site held incredible potential. The goal was to stabilize the soil and prevent more erosion, to redirect stormwater in a way that benefited the landscape, and create a thriving, low-maintenance food forest that could produce

CASE STUDY #3: A FOOD-PRODUCING SHADE GARDEN

This project—my home—sits on a tenth of an acre (less than 6,000 square feet [557 sq m]) on a mostly shaded, sloped lot. I drafted this site analysis when I was doing a second round of planting and a garden overhaul during the COVID-19 pandemic, so many of the fruit trees were planted and the water system had already been installed years prior.

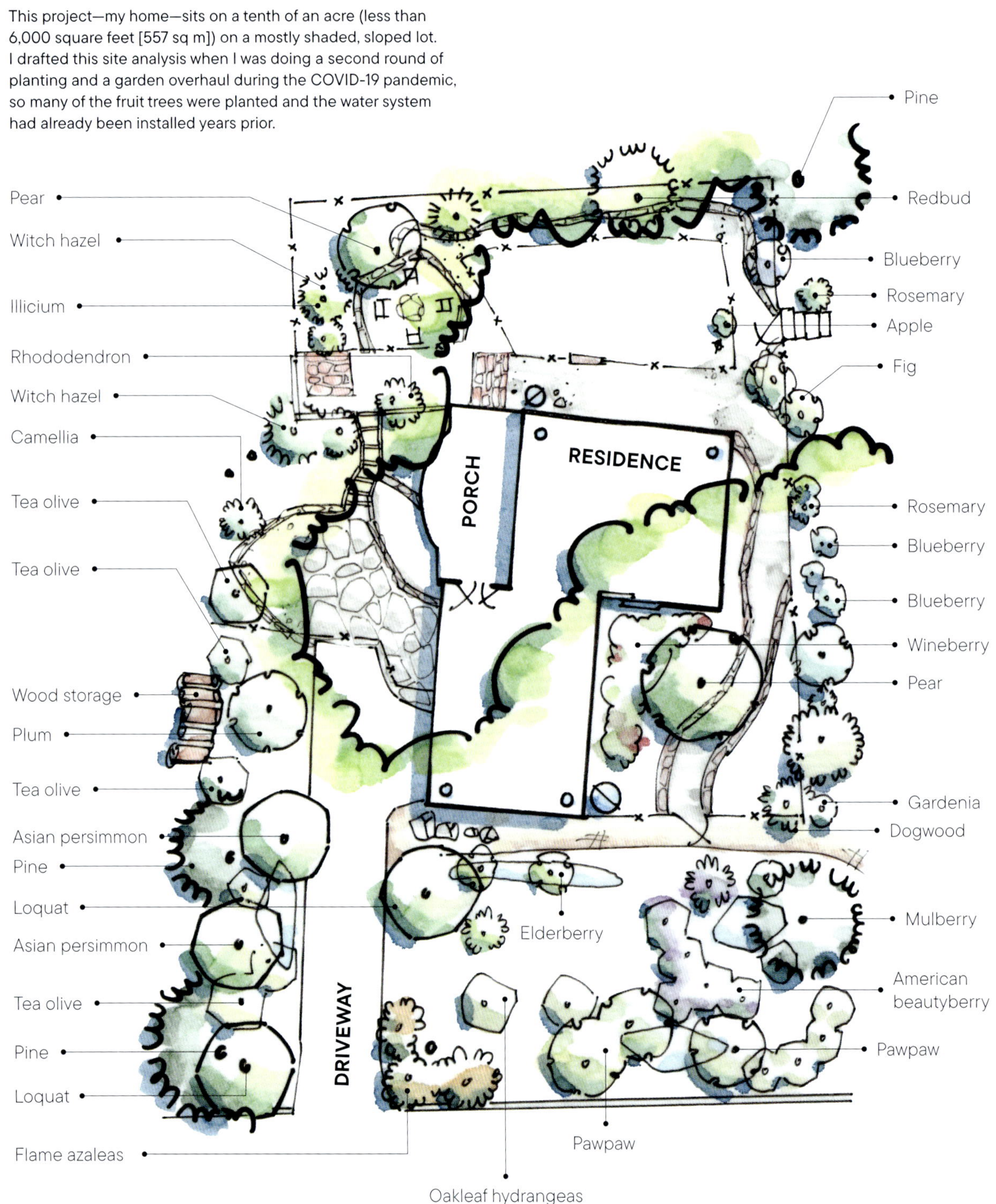

Before

an abundance of fruit, medicinal herbs, and perennial vegetables—even in deep shade.

Rather than relying on fertilizers or constant maintenance, I wanted a self-sustaining landscape where healthy, nutrient-rich soil could develop naturally over time. Beyond water and soil management, I hoped to create a biodiverse environment that supports pollinators, beneficial wildlife, and soil microbes. Instead of a conventional lawn, I wanted an abundance of edible and medicinal perennials that would require minimal upkeep.

Ultimately, my vision was to design a space that felt alive and abundant, solving the site challenges while creating lots of space for my family to enjoy being outside, growing food, and hosting gatherings. I wanted my daughter to know the plants by name and be actively involved in their cultivation and harvest.

Top Years of unchecked runoff had stripped away topsoil, leaving behind compacted clay with very little organic matter.

Bottom The north slope was completely overtaken by English ivy (*Hedera helix*), a fast-spreading aggressive species notorious or suppressing native plants and attracting mosquitoes.

THE SOLUTION

Restoring the Water Cycle

As we've discussed, in conventional landscaping, stormwater is often treated as a problem to be eliminated rather than a resource to be harnessed. Many builders meet stormwater regulations by digging deep, gravel-filled pits and routing roof water into them. While this approach meets requirements, it often damages tree root systems and does little to replenish soil moisture.

Instead, we took a different approach, starting at the highest point of the watershed and working downward to slow, spread, and sink rainwater into the soil. We installed a three hundred-gallon (1,136 L) cistern to capture runoff from the home's thousand-square-foot (93 sq m) roof, providing a backup water source for irrigation during dry spells. The overflow from this tank feeds into a series of rain gardens planted with native and/or medicinal plants such as boneset (*Eupatorium perfoliatum*), doghobble (*Leucothoe fontanesiana*), and rugosa rose (*Rosa rugosa*), which thrive in the alternating wet and dry conditions of a rain garden.

To address foundation drainage issues, we installed subsurface drains that redirect excess water to contour swales along the north-facing slope, formerly covered in English ivy. These swales now support a shade-adapted food forest, featuring a diverse mix of fruiting trees and shrubs, including pawpaw (*Asimina triloba*), loquat (*Eriobotrya japonica*), Asian persimmon

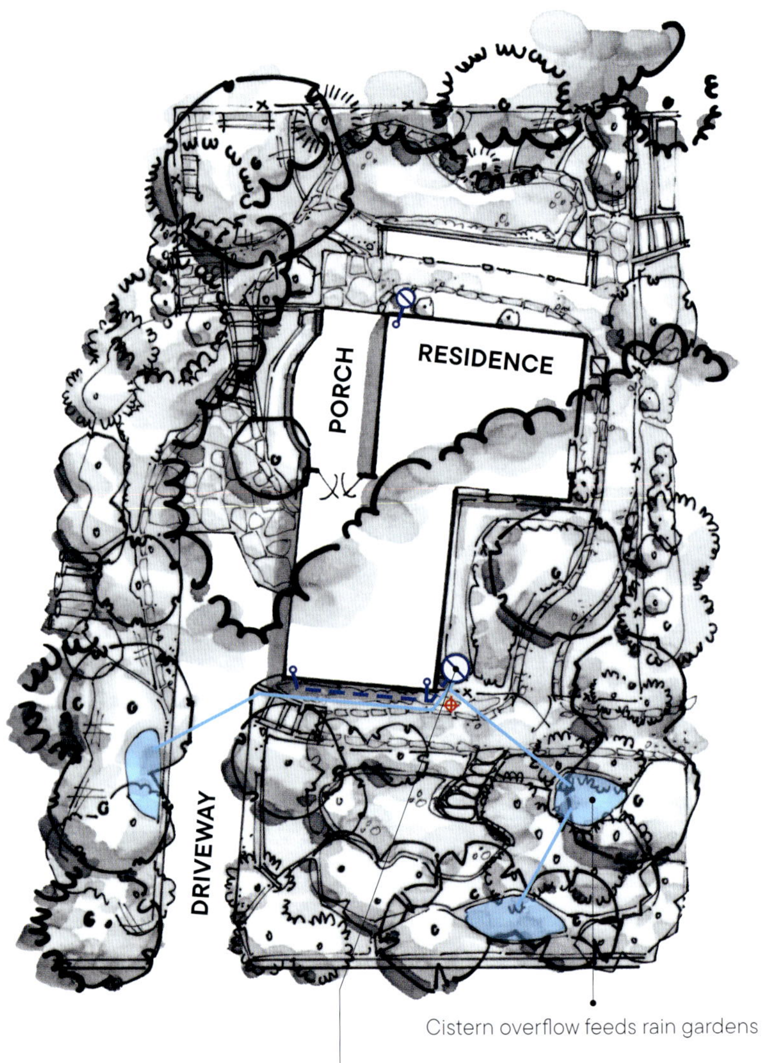

Water from the downspouts is directed to a 300-gallon (1,136 L) cistern, which feeds a single hydrant for watering the small veggie garden. The tank overflows to a series of rain gardens downhill, that grow lots of edible fruits, such as pawpaw (*Asimina triloba*) and mulberry (*Morus rubra*).

After

Top, left We installed a 300-gallon (1,136 L) cistern to capture runoff from the home's 1,000-square-foot (93 sq m) roof.

Top, right Stoke's aster (*Stokesia laevis*) is a drought-tolerant perennial producing large, showy blue to purple flowers from late spring through summer, attracting bees, butterflies, and other pollinators. Its deep root system helps with erosion control.

Right We sought to maximize the space by adding vertical trellises growing a variety of fruiting vines, such as maypop (*Passiflora incarnata*) and hardy kiwi (*Actinidia arguta*).

CASE STUDY #3: A FOOD-PRODUCING SHADE GARDEN

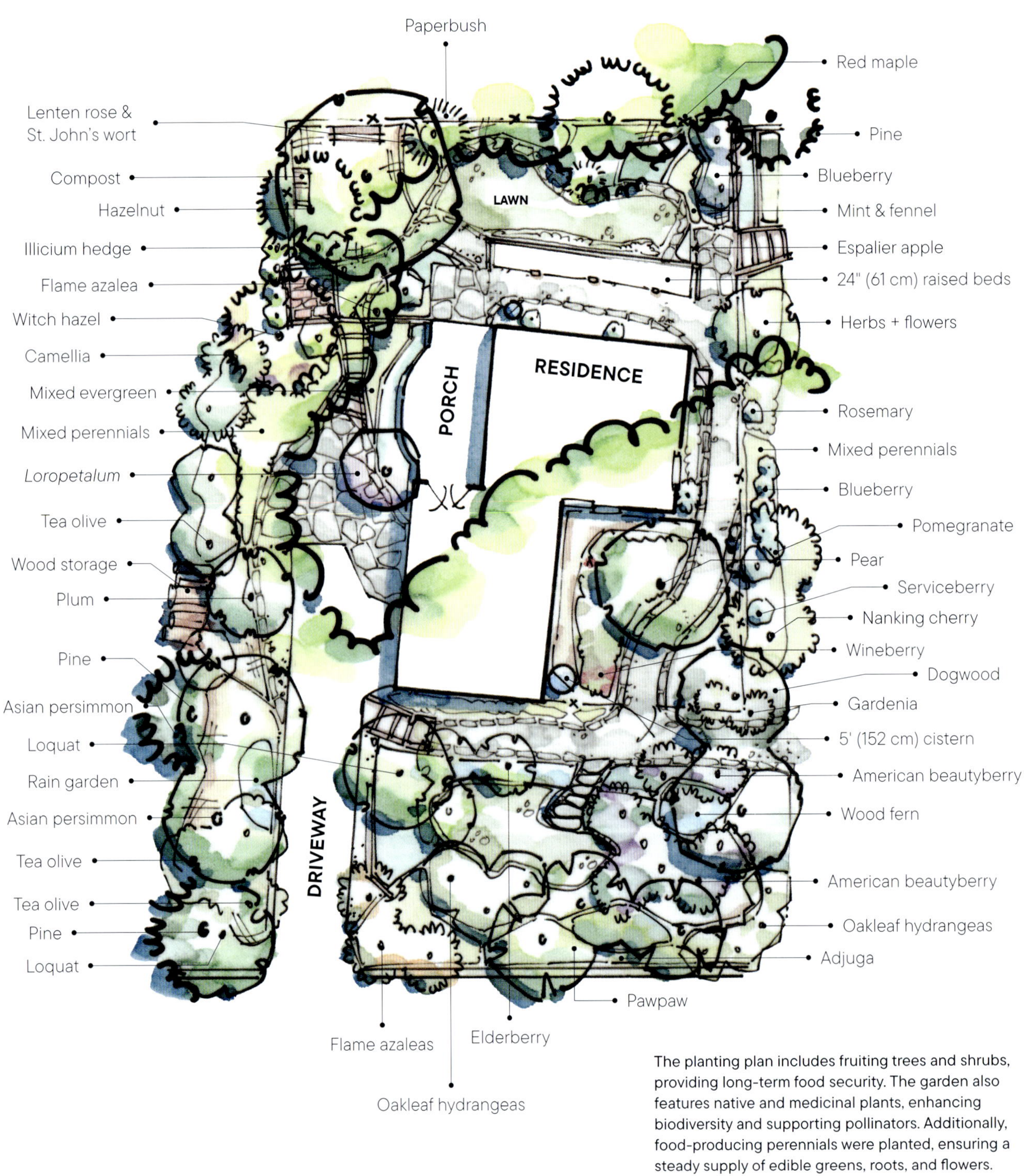

The planting plan includes fruiting trees and shrubs, providing long-term food security. The garden also features native and medicinal plants, enhancing biodiversity and supporting pollinators. Additionally, food-producing perennials were planted, ensuring a steady supply of edible greens, roots, and flowers.

Top Blueberry, *Vaccinium virgatum*, is native to the southeastern United States, and thrives in well-drained, acidic soils and dappled shade. It provides an abundant food source for pollinators, birds, and wildlife, while also being a hardy, drought-tolerant fruiting shrub for edible landscapes with gorgeous fall foliage.

Bottom Creating pervious pathways helped us sink water into the soil. We keep nutrients on-site by using the leaves to mulch garden beds in the autumn. We have a small low-mow, polyculture lawn for open playspace for our daughter.

(*Diospyros kaki*), mulberry (*Morus* spp.), hardy kiwi (*Actinidia arguta*), wineberry (*Rubus phoenicolasius*), elderberry (*Sambucus canadensis*), and American hazelnut (*Corylus americana*). The ground layer includes edible ferns such as ostrich fern (*Matteuccia struthiopteris*), as well as hostas (*Hosta* spp.), Solomon's seal (*Polygonatum biflorum*), and sochan (*Rudbeckia laciniata*).

With these strategies in place, the site now retains and infiltrates approximately 103,650 gallons (392,358 L) of rainwater annually, ensuring the garden remains hydrated while preventing erosion.

Building Soil Fertility

Soil compaction is a widespread issue in urban and suburban landscapes, often exacerbated by conventional construction practices that strip away topsoil and compact subsoil with heavy machinery. Additionally, in cities like Atlanta, many homeowners unknowingly remove valuable organic matter from their landscape by raking and bagging leaves in the fall—only to purchase fertilizer in the spring to replace the lost nutrients.

To rebuild soil fertility naturally, we focused on keeping all organic material on-site. The first step was removing the English ivy, which had been playing its own role in accumulating organic matter. Once the invasive ground cover was cleared, we implemented cover cropping to break up compaction, using daikon radish, Austrian winter pea, and buckwheat in different seasons. These cover crops were chopped and left in place, layered with straw and cardboard, and reseeded for several cycles.

By the time we planted the final perimeter beds, the soil had transformed from compacted, lifeless clay into rich, dark, organic matter-filled earth capable of retaining moisture and supporting plant life with minimal additional inputs.

Cultivating a Thriving Plant Community

Traditional landscape design often prioritizes ornamental plants that are easy to mass-produce and sell year-round. However, these plants are often poorly adapted to their sites, requiring heavy inputs of water, fertilizer, and pesticides to survive. Lawns, in particular, are a major environmental burden, consuming vast amounts of water and fossil fuels for mowing.

Instead, we designed this garden with resilient, multifunctional plants that require minimal maintenance while offering food, medicine, and ecological benefits. The planting plan includes fruiting trees and shrubs, providing long-term food security. The garden also features native and medicinal plants, enhancing biodiversity and supporting pollinators. Additionally, food-producing perennials were planted, ensuring a steady supply of edible greens, roots, and flowers.

For an open play space, we replaced the traditional lawn with a low-maintenance, deep-rooted ground cover mix. The mix includes red fescue (*Festuca rubra*), buffalo grass (*Bouteloua dactyloides*), clover (*Trifolium* spp.), and dandelion (*Taraxacum officinale*), all of which require little upkeep and contribute to soil health. With average rainfall, this

Above American beautyberry (*Callicarpa americana*) is a deciduous shrub producing vibrant purple berry clusters that are an important food source for birds and wildlife while also serving as a natural insect repellent. Its drought tolerance and ability to thrive in diverse soil conditions make it a valuable addition to native and wildlife-friendly landscapes.

Opposite A songbird perches on the fence, attracted to the numerous berries we've planted.

microlawn does not need irrigation and is only mown once a summer with a rotary mower.

A MODEL FOR URBAN LAND STEWARDSHIP

By working with, rather than against, the natural processes at play, this tiny urban homestead has become a pocket-sized powerhouse of productivity and biodiversity. The once-compacted, erosion-prone landscape now captures and stores water, builds its own soil fertility, and provides abundant food and habitat for both people and wildlife.

Urban environments often feel disconnected from nature, but projects like this prove that small spaces can have a big impact. Through thoughtful land stewardship, even a shaded backyard in the middle of a city can contribute to food security, water resilience, and ecological restoration. By making small, intentional choices about how we manage our landscapes, we can all play a role in the healing of our planet.

BUILDING AN OUTDOOR SHOWER THAT'S PRACTICAL, AFFORDABLE, AND ECO-FRIENDLY

For a lot of people, an outdoor shower is a luxury they associate with tropical resorts or dreamy getaways—one of those features you imagine but never think to build at home. But what if I told you that an outdoor shower doesn't have to be expensive or complicated? In fact, it can be a practical, budget-friendly, and eco-conscious addition to your backyard.

At my home in Atlanta, we built an outdoor shower that fits right into our regenerative landscape design. It's more than just a fun feature—it serves a real function for my family, helps hydrate the plants in our yard, and was made mostly from materials we repurposed from past projects.

WHY AN OUTDOOR SHOWER?

Instead of tracking dirt through the house, we wanted a simple way to clean up outdoors.

At the same time, we had an area in our yard—home to a mix of shade-loving plants—that could benefit from a little extra moisture. By designing our shower to drain toward this plant community, we turned what would otherwise be wastewater into a resource.

The shower also adds a little everyday luxury to our space. There's something special about taking a shower outdoors.

PLANNING THE SHOWER LOCATION

The first step was deciding where the shower would go. We needed a spot that was private but still easy to access. It also needed to be close to our home's plumbing, since we wanted hot water, and we wanted to direct the runoff to an area where plants could use it.

We ended up placing the shower near a shaded, wooded slope where we had planted a mix of native species, including rhododendron (*Rhododendron* spp.), flame azalea (*Rhododendron calendulaceum*), and witch hazel (*Hamamelis virginiana*). Since installing the shower, we've added even more moisture-loving plants, like muscadine (*Vitis rotundifolia*), wild hydrangea (*Hydrangea arborescens*), and purple anise (*Illicium floridanum*), all of which thrive with the extra water.

WATER SUPPLY: RAINWATER VS. MUNICIPAL WATER

One of the most frequent questions

We built an outdoor shower that fits right into our regenerative landscape design. It's more than just a fun feature—it serves a real function for my family, helps hydrate the plants in our yard, and was made mostly from materials we repurposed from past projects.

I get is whether our outdoor shower is fed by rainwater. The short answer is no, and here's why:

Our rainwater catchment system is located too far from the shower site, and adding filtration for everyday bathing would have been an expensive extra step. Since we wanted hot water and already had plumbing nearby, it made the most sense to tap into our municipal water supply instead. That said, for anyone looking to build an outdoor shower, rainwater is absolutely an

The shower adds a little bit of luxury to our space. Equipped with hot water and electricity, the shower is useful during most months of the year.

option if you design a system to filter and pressurize it properly.

GATHERING MATERIALS

As with all our home projects, we wanted to use as many salvaged materials as possible. Luckily, we had leftover supplies from previous landscaping and home improvement projects, including scrap wood, bricks, gravel, and some PEX piping.

For the shower's main structure, we repurposed 4 × 4-inch (10 × 10 cm) posts from an old fence. The flooring was made using a mix of salvaged bricks, stone, and pavers set in a bed of pea gravel. To keep costs low, we also sourced extra materials from community forums like Nextdoor and Facebook Marketplace, as well as local stores like Habitat for Humanity ReStore.

The only thing we really splurged on was the brass shower fixture, which we felt was worth the upgrade since we knew we'd be using it almost every day.

BUILDING THE SHOWER

The construction process was surprisingly straightforward. We started by preparing a level base using twelve inches (30 cm) of #57 gravel, which ensures proper drainage and allows the water to saturate the soil and hydrate the plants downslope. Then, we installed four corner posts that we repurposed from arbors and fencing in the yard, securing them in place with concrete footers.

For privacy, we wrapped three sides of the structure with bamboo matting and planted a hedgerow of Florida star anise (*Illicium floridanum*) to create a natural screen. The flooring was made by setting salvaged bricks and pavers on a sand base, then filling in the joints with pea gravel.

The final step was installing the vaulted ceiling, which was by far the trickiest part of the project. We wanted to suspend a light from the ceiling, so we wired an outdoor outlet to one of the posts and hung a linen shower curtain to complete the space.

The whole build took about a day to complete, with some breaks in between to let the posts set and, of course, to manage parenting duties.

EVERYDAY USE AND BENEFITS

Now that the shower is finished, my family uses it almost every day, except in the coldest winter months. It's been one of the best decisions we've made for our outdoor space. Not only is it a great way to rinse off after working outside, but it also doubles as our dog-washing station—an absolute necessity with Georgia's red clay!

Since all the water drains toward our shaded plant community, it also helps keep those moisture-loving plants hydrated, reducing our need for additional irrigation. Of course, we make sure to use biodegradable, plant-safe soaps to keep everything healthy.

CONCLUSION

REGENERATING THE EARTH, ONE YARD AT A TIME

At the heart of this book is a simple but profound belief: Our relationship with the land matters. Whether we live on a rural homestead, a suburban lot, or a city block, the way we tend our surroundings has an impact. Permaculture is not just about growing food or reducing waste; it is about remembering that we are part of the natural world, not separate from it.

I wrote this book because I've seen firsthand how small changes in our landscapes can lead to big shifts in our communities, in our ecosystems, and in our sense of belonging to the land we inhabit. Over the years, I have watched yards transform from high-maintenance, barren spaces into thriving gardens filled with food, pollinators, and life. I have seen the joy in people's eyes when they harvest their first homegrown meal, notice new birds visiting their yard, or realize that their rain garden is helping to restore the local watershed.

Permaculture isn't just about growing things—it's about cultivating relationships. And the first step in that process is simply to start observing, listening, and responding to what the land is telling us.

Throughout this book, we have explored the guiding principles that shape every healthy, self-sustaining landscape, and they are the foundation for making meaningful changes in our own backyards.

Managing Water: Treating It as a Resource, Not a Nuisance

Water is life, yet conventional landscaping treats it as a problem to be eliminated as quickly as possible. We've explored how to slow, sink, and spread water using techniques like swales, rain gardens, and permeable surfaces to keep moisture in the soil rather than losing it to runoff. Whether it's redirecting a downspout into a planting bed or capturing rainwater for irrigation, small changes in how we manage water can create healthier soil, reduce erosion, and even recharge groundwater supplies.

If there's one takeaway from this pillar, it's that we can work with water rather than against it—and in doing so, we restore the land's ability to nourish itself.

Building Soil: Creating the Foundation for Life

Healthy soil is the foundation of every thriving ecosystem. Yet so many modern landscapes have been stripped of their fertility through overdevelopment, compaction, and chemical

Below Bumble bees, now in decline due to habitat loss and pesticides, thrive on native plants like echinacea. Its sturdy blooms provide rich nectar and pollen, making it a vital lifeline for these essential pollinators.

treatments. We've explored simple, regenerative practices like composting, mulching, and cover cropping that rebuild soil health naturally.

By shifting from extractive practices to ones that feed the soil rather than deplete it, we create gardens that require less maintenance, less irrigation, and fewer external inputs.

Cultivating Plant Communities: Working with Nature's Design

As we've seen, conventional landscaping prioritizes neatness and aesthetics over function, often favoring resource-intensive lawns and ornamental plants that require constant upkeep. In contrast, permaculture encourages us to design with biodiversity in mind, choosing plants that not only look beautiful but also provide food, habitat, and ecological benefits.

We've discussed the importance of native plants, food forests, and polycultures, all of which help to create resilient, self-sustaining landscapes. When we plant with intention—choosing species that support soil life, attract pollinators, and thrive in our specific conditions—we create ecosystems that nourish both people and the planet.

A CALL TO ACTION: START WHERE YOU ARE

For me, this journey isn't just professional—it's personal. I came to this work through my own deep love of landscapes, shaped by my childhood experiences in both wild and cultivated spaces. I spent my early years between my mother and stepfather's nursery in Florida, where I saw firsthand the consequences of conventional agriculture, and the mountains of North Carolina, where I witnessed the resilience of intact ecosystems.

Over time, I realized that our landscapes don't have to be extractive; they can be regenerative. We don't have to choose between beauty and function, between human needs and ecological health. A well-designed yard can do both—it can feed people, filter water, sequester carbon, and provide habitat, all while being a space we love to be in.

What I have learned is this: **Permaculture is for everyone**. You don't need a large property, a farming background, or even a green thumb. You just need the willingness to observe, to experiment, and to engage with your space in a new way.

If you take one thing from this book, let it be this: **Your yard matters**. The decisions you make—what you plant, how you manage water, how you care for the soil—have an impact far beyond your property line. Every rain garden, every compost pile, every native plant contributes to a larger movement of land restoration and ecological healing.

The beauty of this work is that it doesn't require perfection, and it doesn't require starting over. You don't have to have all the answers before you begin. Start with observation. Start with small changes. Let your landscape evolve over time, and trust that even the smallest steps are making a difference.

So go outside. Watch where the water flows, feel the soil in your hands, notice what's growing. Start there. The path to regeneration begins at your own front door.

REFERENCES

INTRODUCTION

Atlanta Regional Commission. "Atlanta Region 2024 Population Estimates." Accessed January 29, 2025. atlantaregional.org/what-we-do/research-and-data/atlanta-region-population-estimates.

United Nations Department of Economic and Social Affairs. "68% of the World Population Projected to Live in Urban Areas by 2050, Says UN." accessed January 27, 2025. https://www.un.org/uk/desa/68-world-population-projected-live-urban-areas-2050-says-un.

Audubon, John James. *Ornithological Biography or An Account of the Habits of the Birds of the United States of America.* Vol. 1. A. Black, 1835.

CHAPTER 1

Weiss, Zach. *Hope in a world of crisis: Water cycle restoration.* TEDx Talks, June 3, 2019. https://www.ted.com/talks/zach_weiss_hope_in_a_world_of_crisis_water_cycle_restoration.

Makarieva, Anastassia M., and Victor G. Gorshkov. "The Biotic Pump: Condensation, Atmospheric Dynamics and Climate." *International Journal of Water* 5, no. 4 (2010): 365 -385. https://doi.org/10.1504/IJW.2010.038729.

Working Group I Contribution to the Sixth Assessment Report of the Intergovernmental Panel on Climate Change,. *Climate Change 2021: The Physical Science Basis.* (Intergovernmental Report on Climate Change, 2021). ipcc.ch/report/ar6/wg1/.

Ask Extension: What Percentage of the Mass of a Tree is Water? (General Questions). (n.d.). Illinois Extension. Accessed June 22, 2023. web.extension.illinois.edu/askextension/thisQuestion.cfm?ThreadID=19549&catID=192&AskSiteID=87

U.S. Environmental Protection Agency. "Statistics and Facts." *WaterSense.* Updated March 24, 2025. epa.gov/watersense/statistics-and-facts.

Eagleson, P. S. "The Role of Water in Climate." *Proceedings of the American Philosophical Society* 144, no. 1 (2000): 33-38. jstor.org/stable/1515603.

CHAPTER 2

Briedè, J.-W. "Deep Thoughts about Stormwater Runoff." *Plant More Plants* blog, October 20, 2011. Accessed June 22, 2023. http://www.plantmoreplants.com/blog/deep-thoughts-about-stormwater-runoff.

Lancaster, Brad. *Rainwater Harvesting for Drylands and Beyond: Guiding Principles to Welcome Rain into Your Life and Landscape.* TBS/GBS/Transworld, 2019.

Ang, Carmen. "The Median Home Size in Every U.S. State in 2022." Visual Capitalist, November 22, 2022. Accessed June 23, 2023. https://www.visualcapitalist.com/cp/median-home-size-every-american-state-2022/.

CHAPTER 4

Hopkinson, Jenny. "Can American Soil Be Brought Back to Life?" *POLITICO*, September 13, 2017. Accessed January 29, 2025. politico.com/agenda/story/2017/09/13/soil-health-agriculture-trend-usda-000513/.

Schwartz, Judith D. "Soil as Carbon Storehouse: New Weapon in Climate Fight?" *Yale Environment* 360, (March 4, 2014). Accessed January 30, 2025. https://e360.yale.edu/features/soil_as_carbon_storehouse_new_weapon_in_climate_fight.

Mulvaney, R. L., S. A. Khan, and T. R. Ellsworth. "Synthetic Nitrogen Fertilizers Deplete Soil Nitrogen: A Global Dilemma for Sustainable Cereal Production." *Journal of Environmental Quality* 38, no. 6 (2009): 2295–2314. Accessed January 31, 2025. doi.org/10.2134/jeq2008.0527.

Virginia Institute of Marine Science. "Dead Zones Continue to Spread." August 14, 2008. Accessed January 31, 2025. www.vims.edu/newsandevents/topstories/archives/2008/dead_zones_spread.php

Erickson, Jim. "Researchers Predict Average 'Dead Zone' for Gulf of Mexico in 2016." *Phys.org*, June 10, 2016. Accessed January 31, 2025. phys.org/news/2016-06-average-dead-zone-gulf-mexico.html.

U.S. Environmental Protection Agency. "Reduce Your Outdoor Water Use." Accessed February 14, 2025. nepis.epa.gov/Exe/ZyPURL.cgi?Dockey=P100R9D8.TXT.

U.S. Environmental Protection Agency. *Municipal Solid Waste Generation, Recycling, and Disposal in the United States: Facts and Figures for 2011.* (U.S. EPA, 2013), archive.epa.gov/epawaste/nonhaz/municipal/web/pdf/mswcharacterization_508_053113_fs.pdf

Minnesota Association of Soil and Water Conservation Districts. *"Our Soil: A Layer of Life" Study Guide.* Accessed February 14, 2025. www.maswcd.org/Youth_Education/StudyGuides/Soils_study_guide.htm.

CHAPTER 5

Lal, Rattan. "Soil Carbon Management and Climate Change." *Carbon Management* 4, no. 4 (2013): 439–462. Accessed January 31, 2025. https://doi.org/10.4155/cmt.13.31.

CHAPTER 7

Zahm, S. H., and M. H. Ward. "Pesticides and Childhood Cancer." *Environmental Health Perspectives* 106, no. Suppl 3 (1998): 893–908.

Beyond Pesticides. "Lawn Pesticide Facts and Figures." August 2005. Accessed February 5, 2025. beyondpesticides.org/assets/media/documents/lawn/factsheets/LAWNFACTS%26FIGURES_8_05.pdf.

Mulvaney, R. L., Khan, S. A., and Ellsworth, T. R. "Synthetic Nitrogen Fertilizers Deplete Soil Nitrogen: A Global Dilemma for Sustainable Cereal Production." *Journal of Environmental Quality* 38, no. 6 (2009): 2295–2314. doi.org/10.2134/jeq2008.0527.

Diaz, Robert J., and Rutger Rosenberg. "Spreading Dead Zones and Consequences for Marine Ecosystems." *Science* 321, no. 5891 (2008): 926–929. science.org/doi/10.1126/science.1156401.

National Oceanic and Atmospheric Administration. "NOAA forecasts average-sized 'dead zone' for the Gulf of Mexico." June 3, 2021. Accessed February 5, 2025. noaa.gov/news-release/noaa-forecasts-average-sized-dead-zone-for-gulf-of-mexico.

Warming, Johannes Eugenius Bülow. *Plantesamfund: Grundtræk af den økologiske Plantegeografi.* Copenhagen: H. Hagerup, 1895, summarized in Pedar Anker, Chapter 23 of *Ecology Revisited: Reflecting on Concepts, Advancing Science* (Springer Science & Business Media, 2011), 325-331. https://pederanker.com/wp-content/uploads/2011/06/plant-community-ecology-revisited.pdf.
Wikipedia, "Clements, Frederic," last modified 20 January 2025, at 02:37 (UTC) en.wikipedia.org/wiki/Frederic_Clements.
Eurostat. "Agri-environmental Indicator - Consumption of Pesticides." Accessed February 14, 2025. ec.europa.eu/eurostat/statistics-explained/index.php/Agri-environmental_indicator_-_consumption_of_pesticides.
European Environment Agency. "How Pesticides Impact Human Health." Accessed February 14, 2025. eea.europa.eu/publications/how-pesticides-impact-human-health.
Economic Research Service. "Lawn and Garden Industry Overview." United States Department of Agriculture. Accessed February 14, 2025. https://www.ers.usda.gov/.
NASA Earth Observatory. "Lawn Surface Area in the United States." National Aeronautics and Space Administration. Published June 29, 2005. Accessed February 14, 2025. earthobservatory.nasa.gov/images/6019/lawn-surface-area-in-the-united-states.
Milesi, Cristina, et al. "Mapping and Modeling the Biogeochemical Cycling of Turf Grasses in the United States." *Environmental Management* 36, no. 3 (2005): 426–438. link.springer.com/article/10.1007/s00267-004-0316-2.
McHugh, Margaret Anne. "Lawns Are an Outdated Cultural Norm. Let's Lose Them—Before We Lose the Pollinators." *Toronto Star*, June 26, 2022. Accessed June 12, 2025. https://www.thestar.com/opinion/contributors/lawns-are-an-outdated-cultural-norm-let-s-lose-them-before-we-lose-the-pollinators/article_da15d992-7d54-5fc3-bee8-9f00d512396d.html.
Eurostat. "Land Cover Statistics." *Statistics Explained.* European Commission. Last modified June 2021. Accessed February 14, 2025. https://ec.europa.eu/eurostat/statistics-explained/index.php/Land_cover_statistics.
Wang, Ya, et al. "The Ecohydrological Cost of Lawns." *American Geophysical Union, Fall Meeting 2018*, abstract B33G-2752. ui.adsabs.harvard.edu/abs/2018AGUFM.B33G2752P/abstract.
Law, Beverly E., et al. "Carbon Sequestration and Greenhouse Gas Emissions in Urban Turf." Environmental Research Letters 9, no. 7 (2014): 074003. iopscience.iop.org/article/10.1088/1748-9326/9/7/074003.
Townsend-Small, Amy, Czimczik, Claudia. "Carbon Sequestration and Greenhouse Gas Emissions in Urban Turf." *Geophysical Research Letters* 37, no. 2 (2010): agupubs.onlinelibrary.wiley.com/doi/full/10.1029/2009GL041675
Guo, Xinying, et al. "Beyond Agriculture: Land Use Thresholds Governing Pesticide Mixture Risks in Megacity Surface Waters." *Journal of Hazardous Materials.* https://doi.org/10.1016/j.jhazmat.2025.138657.

PHOTOGRAPHY CREDITS

Audra Melton: pages 4, 9 (top left), 15 (right), 20, 46, 186, 187 (left), 196
Ash Wilson: pages 11 (left), 41, 42, 52, 55, 61 (right), 63, 136, 152, 155, 156
Erik Meadows: pages 10, 12 (left), 15 (middle), 17, 28, 30–31 (middle), 36, 37 (bottom), 44, 61 (left), 68, 74, 75, 85–89, 93, 104, 116, 121, 124–128, 131, 132, 138, 144–146, 149, 154, 157, 159, 162, 168–174, 177 (bottom), 178, 179, 180–182, 184, 191 (top right and bottom), 193, 195, 199
Jonathan Banks: pages 7, 9 (bottom left), 11 (right), 12 (right), 19, 34 (left), 45, 47, 48, 49, 62, 67, 82–83, 94, 95 (center and middle), 98, 110, 112, 115, 120, 123, 130, 140–143, 150 (middle), 202
Sara Callaway: page 187 (right)
Shades of Green Permaculture: pages 9 (top right and bottom right), 11 (middle), 15 (left), 16, 54, 81, 92, 96, 99, 100, 111, 150 (left), 161 (top), 165, 175, 177 (top), 189
Shutterstock: pages 22, 24–27, 31 (top right), 34 (right), 35, 50, 76, 78, 90, 91, 95 (right), 102, 118, 122, 148, 158, 161 (bottom)
Virginie Kipellen: page 197
Wildgrain Photography: pages 37 (top), 40, 71, 150 (right), 194, 191 (top left)

ABOUT THE AUTHOR

Brandy M. Hall is the founder and CEO of Shades of Green Permaculture, a visionary leader in regenerative design dedicated to transforming landscapes into thriving ecosystems. With a background in ecological restoration, permaculture, and sustainable water management, Brandy has spent over two decades helping homeowners, businesses, and communities create landscapes that nourish people and the planet. A passionate educator and advocate for environmental resilience, she blends science, artistry, and deep ecological knowledge to craft regenerative solutions that restore biodiversity, conserve resources, and build climate resilience.

In addition to her work in permaculture, Brandy serves as mayor of a small city in the metro-Atlanta area, using her platform to advocate for policies that support sustainable land use, ecological restoration, and resilient local communities. She believes that regenerative solutions extend beyond landscapes and into governance, where thoughtful leadership can create lasting positive change.

Brandy's work has earned national recognition, including features in *Forbes*, NPR, and *The New York Times*, as well as invitations to speak at leading sustainability conferences and educational institutions.

At the heart of her work is a deep commitment to her family and community, recognizing that strong relationships—both with people and the land—are the foundation of a truly regenerative future. Whether designing landscapes, shaping policy, or spending time with her family outside, Brandy is dedicated to fostering a world where future generations can thrive.

ACKNOWLEDGMENTS

This book would not exist without the love, support, and wisdom of the people who've shaped me and this work. I stand on the shoulders of many incredible teachers—some I've sought out, some I was lucky enough to be born into, and others who appeared at just the right time.

Zephyr: Your wonder, curiosity, and pure-hearted belief that the world is good keep me working to make it so. You are courageously yourself, and that authenticity is a seed I will always protect and nurture. You remind me what this work is for.

Aaron: Your quiet devotion to our Little Dove has been the fertile soil from which this book has grown. Your craftsmanship—of both language and life—embodies the regenerative values at the heart of this work. You remind me to build beauty and believe in what we can't yet see.

Mom: Your tireless work ethic and orientation toward service taught me to meet the needs of the moment with courage and humility. You modeled how to care deeply and act decisively—skills this work demands every day.

Dad: Your irreverent humor and love of rivers and ridgelines taught me how to be in relationship with the wild world, and to never take myself too seriously while doing serious work.

Mark: Your generosity and love of travel opened my eyes early to the value of sharing and cultural exchange. You led me to Central America, where I first encountered permaculture and began unlearning the comfort of certainty.

Kathy: Your sharp thinking and willingness to ask hard questions have kept me honest. You push me to align my values with my actions, again and again.

Joyce, Laurel, and Christopher: Thank you for showing up in the ways that matter. Christopher, I miss your laugh every day. It reminds me to savor this fleeting life; tomorrow is not a given.

Jonathan: You've witnessed and supported me through one of the hardest periods of my life while I wrote this book. You remind me to reroot in my trust of the Creative daily, and know that the earth supports me. Thank you for being a lighthouse.

Terry: You saw a writer in me from such a young age, and you helped cultivate my love of language.

A heartfelt thank you to Ben, for teaching me "everything I know," and to Rebecca, a grounding force and my right hand for the better part of a decade, for your keen insights, thoughtful critiques, and dedication to refining the concepts in this book into what they are today. And to the entire Shades of Green team: Thank you for living these principles, for showing me what's possible when we design with care, and for walking this journey day in and day out. A huge thank you to all of our clients over the years who have trusted us to bring your landscape to life, and have given us the grace to learn in the process.

To all my friends—you know who you are: Thank you for listening and encouraging me, and reminding me of my own resilience.

Finally, to you, dear reader: Thank you for engaging with this work. Whether you're just beginning or deep into your permaculture journey, I hope this book offers tools, insights, and inspiration to nourish your path. May it be a companion as you reimagine your relationship with the land, your community, and yourself.

INDEX

Page numbers in italics indicate a photograph.

F

G

H

I

Q

R

S